ecofeminism

ecofeminism

feminist intersections with other animals and the earth

Edited by Carol J. Adams and Lori Gruen

Bloomsbury Academic
An imprint of Bloomsbury Publishing Plc

B L O O M S B U R Y
NEW YORK • LONDON • NEW DELHI • SYDNEY

Bloomsbury Academic

An imprint of Bloomsbury Publishing Inc

1385 Broadway	50 Bedford Square
New York	London
NY 10018	WC1B 3DP
USA	UK

www.bloomsbury.com

BLOOMSBURY and the Diana logo are trademarks of Bloomsbury Publishing Plc

First published 2014
Reprinted 2015 (twice)

Library of Congress Cataloging-in-Publication Data

A catalog record for this book is available from the Library of Congress.

ISBN: HB: 978-1-6235-6590-9
PB: 978-1-6289-2803-7
ePub: 978-1-6289-2622-4
ePDF: 978-1-6289-2197-7

Typeset by Fakenham Prepress Solutions, Fakenham, Norfolk NR21 8NN
Printed and bound in the United States of America

Wrenching an ethical problem out of its embedded context severs the problem from its roots … In a sense, we are given truncated stories and then asked what we think the ending should be. However, if we do not understand the worldview that produced the dilemma that we are asked to consider, we have no way of evaluating the situation except on its own terms.

Re-specting nature literally involves "looking again." We cannot attend to the quality of relations that we engage in unless we know the details that surround our actions and relations. If ecofeminists are sincere in their desire to live in a world of peace and nonviolence for all living beings, we must help each other through the pains-taking process of piecing together the fragmented worldview that we have inherited. But the pieces cannot simply be patched together. What is needed is a reweaving of all the old stories and narratives into a multifaceted tapestry.

Marti Kheel 1993

Contents

Acknowledgments

After the death of our dear friend Marti Kheel on November 19, 2011, a community of mourners came together online and at memorials on both coasts in an embrace of compassion and care that lifted us up personally and enlivened us politically to share the insights of ecofeminism. We are deeply grateful to this extended community and to the Kheel family for facilitating this renewal of ecofeminist theory and practice in the wake of Marti's death.

To build on the conversations that were happening, we organized a conference "Finding a Niche for All Animals" at Wesleyan University in 2012. We are grateful to all of the participants at the conference for sharing their recollections, their new ideas, and their excitement about making the world better for animals and the earth. Some of the papers presented at the conference are reworked in the chapters in this volume and we thank those contributors as well as those scholars and activists whose papers were not ultimately included. We are grateful to the sponsors of the conference: Wesleyan Animal Studies, the College of the Environment, the Ethics in Society Project, the Feminist, Gender, and Sexuality Studies Program, the Center for the Study of Public Life, and the Philosophy Department at Wesleyan and the Animals and Society Institute, Feminists for Animal Rights, Arnold S. and Ellen Kheel Jacobs, AJ Jacobs, Jane Kheel Stanley, and other members of the Kheel family. Lynn Higgs and Hilda Vargas provided outstanding logistical support for which we are so grateful. The conference would not have happened without the unwavering support of Jane Stanley and Batya Bauman, who also created a new website for Feminists for Animal Rights: http://www.farinc.org/ after the conference.

We thank three anonymous reviewers of this manuscript for their thoughtful suggestions. We are particularly grateful to Kevin Ohe, Haaris Naqvi, and Laura Murray at Bloomsbury Press for their enthusiasm and encouragement in publishing this book.

The night Marti died, Carol and Lori began a conversation about mourning and remembrance. That night Lori suggested the idea of holding a conference. Carol wishes to acknowledge the roots of this book in that sad night, and how the work of collective mourning allowed for the emergence of many of these important essays; that this book exists is due in great part to Lori's spirit, insights, and skill, as well as her deep understanding of ecofeminist philosophy and activism.

LG and CJA, November 26, 2013

Notes on contributors

Ralph R. Acampora is Associate Professor of Philosophy at Hofstra University. He is the author of *Corporal Compassion: Animal Ethics and Philosophy of Body* (University of Pittsburgh Press, 2006), edited *Metamorphoses of the Zoo: Animal Encounter After Noah* (Lexington Books, 2010) and co-edited *A Nietzschean Bestiary* (Rowman & Littlefield, 2003). Recent interests of his include the hermeneutics of spectatorship at zoos, moral issues pertaining to the built, including biotechnical environment, and the ontological status of nature. A vegetarian who tries to be vegan in an overwhelmingly omnivorous and carno-crazed culture, he does not eat friends, lovers, or close kin (of any species).

Carol J. Adams is the author of *The Sexual Politics of Meat*, now in a twentieth anniversary edition. She edited the first multicultural ecofeminist text focusing on religion, *Ecofeminism and the Sacred*. Her activist work in the 1980s challenging racism and domestic violence and for abortion rights informs her scholarship. She is on the Board of Directors of Minding Animals International. For more information see www.caroljadams.com

Deane Curtin is the Hanson-Peterson Professor of the Liberal Arts and Professor of Philosophy at Gustavus Adolphus College. While on sabbatical writing his chapter in India he coordinated two projects at the request of His Holiness, the Dalai Lama. At the Library of Tibetan Works and Archives in Dharamsala, he coordinated a project to translate the major texts of Western philosophy into Tibetan for the first time. He also designed and taught a core ethics course based on the Dalai Lama's book *Beyond Religion* at the Dalai Lama Institute for Higher Education in Bangalore.

Josephine Donovan, Professor Emerita of English at the University of Maine, is the author or editor of 13 books and numerous articles. With

Carol J. Adams she co-edited *The Feminist Care Tradition in Animal Ethics* (2007), *Beyond Animal Rights* (1996), and *Animals and Women: Feminist Theoretical Explorations* (1995).

Karen S. Emmerman earned her doctorate in philosophy in 2012 from the University of Washington, where she wrote a dissertation in ecofeminist animal ethics, "Beyond the basic/non-basic interests distinction: A feminist approach to inter-species moral conflict and moral repair." Karen is an adjunct lecturer in the UW Philosophy department and Comparative History of Ideas program, co-organizer of the University of Washington Critical Animal Studies Working Group, and board member of the UW Center for Philosophy for Children.

Greta Gaard's research and activism address the local and global intersections of gender, race, sexuality, species, and ecology. Her many published essays bring a feminist perspective to explore intersections of social, species, and environmental justice. Her book publications include *Ecological Politics: Ecofeminists & the Greens* (1998); *Ecofeminism: Women, Animals, Nature* (1993); *Ecofeminist Literary Criticism* (1998); *International Perspectives in Feminist Ecocriticism* (2013), and a book of creative nonfiction, *The Nature of Home* (2007).

Lori Gruen is Professor of Philosophy, Feminist, Gender, and Sexuality Studies, and Environmental Studies at Wesleyan University, where she also coordinates Wesleyan Animal Studies. She has published extensively on topics in ecofeminist ethics, animal ethics, and environmental philosophy. She is the author of two books on animal ethics, most recently *Ethics and Animals: An Introduction* (Cambridge, 2011); the editor of *The Ethics of Captivity* (Oxford, 2014); co-editor of five books; and she co-edited the special issue of *Hypatia* on Animal Others, with Kari Weil. For more information go to lorigruen.com

pattrice jones is a cofounder of VINE Sanctuary, an LGBTQ-run farmed animal sanctuary that works within an extended, ecofeminist understanding of the intersection of oppressions. She is the author of *Aftershock: Confronting Trauma in a Violent World* (Lantern, 2007) and has published essays in *Confronting Animal Exploitation* (McFarland, 2013); *Sister Species* (University of Illinois Press, 2011); *Sistah Vegan* (Lantern, 2010); *Contemporary Anarchist Studies* (Routledge, 2009); *Igniting a Revolution* (AK Press, 2006); and *Terrorists or Freedom Fighters* (Lantern, 2004).

Claire Jean Kim's first book, *Bitter Fruit: The Politics of Black-Korean Conflict in New York City* (Yale University Press, 2000) won two awards from the American Political Science Association. Her second book, *Race, Species and Nature in a Multicultural Age* (Cambridge University Press, 2014), examines the intersection of race, species, and nature in impassioned disputes over how immigrants of color, racialized minorities, and Native people in the US use animals in their cultural traditions. She has written numerous journal articles and book chapters and co-guest edited, with Carla Freccero, a special issue of *American Quarterly* entitled, Species/Race/Sex (September 2013).

Deborah Slicer is Professor of Philosophy at the University of Montana, where she directs the Masters Program in Environmental Philosophy. She has published most recently on the moral value of quiet, on Thoreau, on poetics, and on animals and ethics. Her collection of poetry, *The White Calf Kicks*, won the 2003 Autumn House Poetry Prize, judged by Naomi Shihab Nye.

Sunaura Taylor is an artist, writer, and activist. Taylor's artworks have been exhibited at venues across the country, including the CUE Art Foundation, the Smithsonian Institution and the Berkeley Art Museum. She is the recipient of numerous awards including a Joan Mitchell Foundation MFA Grant and an Animals and Culture Grant. Her written work has been published in the *Monthly Review, Yes! Magazine, American Quarterly* and *Qui Parle*. Her book *Beasts of Burden* (Feminist Press 2014) explores the intersections of animal ethics and disability studies. She is currently a PhD student in American Studies in the Department of Social and Cultural Analysis at NYU.

Richard Twine is a sociologist and is currently a Research Fellow at the Institute of Education, University of London. He was previously at the Universities of Lancaster and Glasgow. His current research explores how to change food habits in the context of climate change. He is author of the book *Animals as Biotechnology—Ethics, Sustainability and Critical Animal Studies* (Routledge, 2010), and editor (with Nik Taylor) of *The Rise of Critical Animal Studies—From the Margins to the Centre* (Routledge, 2014), as well as articles on the animal-industrial complex, antibiotics, de-domestication, ecofeminism, posthumanism, and bioethics.

List of illustrations

Introduction
Carol J. Adams and Lori Gruen

cofeminism addresses the various ways that sexism, hetero-normativity, racism, colonialism, and ableism are informed by and support speciesism and how analyzing the ways these forces intersect can produce less violent, more just practices. In the 1990s, ecofeminists worked to remedy a perceived problem in feminist theory, animal advocacy, and environmentalism, namely, a lack of attention to the intersecting structures of power that reinforce the "othering" of women and animals, and contribute to the increasing destruction of the environment. Though sometimes called "utopian" or "concerned with too many issues," ecofeminist theory exposes and opposes inter-secting forces of oppression, showing how problematic it is when these issues are considered separate from one another. This approach also identifies the shortcomings with mainstream "animal rights" treatments of speciesism.

In large part because of misunderstandings, ecofeminism fell out of favor for a number of years. Now ecofeminism is attracting renewed attention as the impact of human activities on the more-than-human world worsens. Ecofeminist theory helps us imagine healthier relationships; stresses the need to attend to context over universal judgments; and argues for the impor-tance of care as well as justice, emotion as well as rationality, in working to undo the logic of domination and its material and practical implications. This volume deepens these significant insights.

1847. The word "vegetarian" is coined at Ramsgate, England from the Latin "vegetus" meaning "whole, sound, fresh or lively."

Ecofeminism: Feminist Intersections with Other Animals and the Earth emerged from our collective mourning of the loss of one of ecofeminism's foremothers, Marti Kheel, who died in 2011. Her death catalyzed conversations about where ecofeminism has been and where it might go. Noting that the excitement and vigor of early ecofeminist work had been lost to a new generation of activists and scholars and that misinterpretations and misrepresentations dominated the discussion when ecofeminism did come up, we felt it was time to reintroduce the intersectional concerns that are at the heart of ecofeminism. We came together at Wesleyan University in 2012 for a conference celebrating what ecofeminism had accomplished, looking at the creative tensions within activism and scholarship, with the hope of moving ecofeminist theory and practice forward.

The book begins with a historical, grounding section called "Groundwork" that situates ecofeminist theory and activism and presents personal accounts that reflect experiences making connections. "Groundwork" reveals how, even before *Animal Liberation* and *The Case for Animal Rights*—two books linked to the emergence of the animal rights movement and animal studies—feminists were expressing formative theoretical and practical insights. Two sections follow "Groundwork," one on "Affect" in which authors dig deeper into how our affective, emotional connections to other animals and the earth inform our theory and practice. In the next section "Context," authors explore the complexities associated with making decisions while giving full consideration to changing contexts.

The practice of making connections between the oppression of women, people of color, indigenous peoples, workers, and other animals has been going on for a long time. On the right side pages of this book, you will find a timeline that represents our initial efforts to identify intersectional, nonviolent activisms that include nonhuman animals. Our acknowledgment of the work of social justice activists is by no means complete and we hope that others may be inspired by flipping through the history to build on the evolving project of uncovering intersectional anti-speciesist activism through time and across cultures.

It is by now familiar to most people who have thought about the ethical and political grounds for our obligations to the other-than-human world, that reason alone cannot motivate and sustain a rejection of destructive anthropocentric practices. The feminist care tradition in ethics was developed as an alternative to the rights-based justice accounts that had dominated discussions within the academy and in social justice movements. Though many feminists saw "care" as a necessary complement to "justice," the justice/care debate was often framed in binary terms, where our responsibilities and motivations were seen as a matter of justice **or** as a function of our capacities to care. Ecofeminists identify dualistic thinking (that creates inferior others and upholds certain forms of privilege as in the human/animal, man/woman, culture/nature, mind/body dualism) as one of the factors that undergirds oppression and distorts our relationships with the earth and other animals. So the feminist care tradition in animal ethics, critical of dualistic thinking, challenged the reason/emotion binary and the elevation of abstract, universal principles deduced through detached reasoning over particular sympathies and sensitivities to the plight of those whose lives and well-being are in jeopardy.

The feminist care tradition focuses on affective connections, including compassion and empathy, and shows how these connections have a cognitive or rational component. We can make empathetic mistakes or our compassion can be directed in the wrong ways; our deliberative and epistemic capacities help to channel our affective engagements appropriately. Ecofeminist attention to these affective connections resonates with but is not the same as what is now called "affect theory" that began as a view in psychology focused on the neuroscience of emotions and then was taken up by theorists who were drawn to materialist interpretations of experience. By attending to responsive bodies as sites of organized, non-intentional subjectivity, affect theory may prove particularly useful in understanding the agency of other animals and this is important for ecofeminists who are concerned about the dangers of anthropocentric projections of sameness onto others. But ecofeminists want to avoid the dualisms that appear in much of the writing of affect theorists who maintain distinct divisions between the systems of reason and emotion, intention and embodiment, cognition and affect.

There is exciting new territory to explore in thinking about the relationship between affect, ethics, other animals, and the environment. The chapters in the first section examine compassion, sympathy, joy, eros, vulnerability, and grief. The section begins with "Compassion and Being Human" in which Deane Curtin, who originally coined the phrase "contextual moral vegetarianism," explores the relationship of affect and reason as it is expressed in an ethic of compassion.

In the second essay, Deborah Slicer, through a poignant story of a playful horse she befriended, explores the way that joy and laughter transcend species boundaries and create moral connection. Like Slicer, Josephine Donovan believes that our communication with other animals is far more nuanced and accessible than many have thought. Donovan, one of the originators of the feminist care tradition in animal ethics, muses on quantum physics to develop what she calls "participatory epistemology." As reason has had such a prominent role in discussions of both epistemology and ethics, explorations of the nature and force of desire both as a source of

1878. Anti-vivisectionist Frances Power Cobbe writes "Wife Torture in England."

knowing and as a motivation for acting has, until the recent affective turn, been less analyzed.

In her provocative essay, pattrice jones explores the nature of eros in humans and other beings as a pleasurable font of connection. Sunaura Taylor's essay, "Interdependent Animals: A Feminist Disability Ethic-of-Care," challenges the myth of self-sufficiency and the way that those in power have used "dependency" as a rhetorical tool to attempt to justify exploitation. In the last essay in this section, Lori Gruen addresses the difficult issues of grief and mourning. In order to make the lives of marginal others intelligible, Gruen suggests the need for coming to terms with death and dying and recognizing our culpability in those deaths, even while we try to minimize them.

Wary of overgeneralization, ecofeminists have been critical of holist tendencies among environmentalists and the focus on rights among animal advocates. As ecofeminist Marti Kheel argued, the universal standpoint that is commonly invoked within environmental and animal ethics truncates a larger narrative that questions how problems emerge in particular contexts. In the "Context" section, the chapters explore the ways that our ethical, political, and epistemic commitments are challenged when contexts change.

The section starts off with one of the hottest topics within animal activism and animal studies today—when and whether eating other animals is ever justified. In "Caring Cannibals and Human(e) Farming: Testing Contextual Edibility for Speciesism," Ralph Acampora explores just how far arguments for contextual moral veganism can go and remain anti-speciesist. Karen Emmerman takes the tension between context and species further in "Inter-Animal Moral Conflicts and Moral Repair: A Contextualized Ecofeminist Approach in Action." She explores the wrenching conflict she experienced when she gave birth to her son prematurely. As a vegan who is committed to raising her child as a vegan, she was faced with a dilemma while in the hospital: whether to save her child's life by using a product that involved the sacrifice of sheep.

In "The Wonderful, Horrible Life of Michael Vick," Claire Kim explores the ways that race and class are often used to obscure prejudice based on species while simultaneously observing that animal advocates often feign a dangerous "color-blindness." Awareness of context can be an important counter to the universalizing prescriptions that often characterize liberation struggles. Mindful of the ways in which universal demands for vegetarianism, for example, recapitulate colonial, ethnocentric attitudes and practices, many ecofeminists argue for contextual moral veganism, but in his piece, "Ecofeminism and Veganism: Revisiting the Question of Universalism," Richard Twine positively supports the idea that contextual moral veganism may not be that far from universalism after all.

Carol Adams argues that cultural theory must include consideration about species hierarchies and attitudes when examining racial and sexual representations. She shows how critical theory that analyzes the "reclining nude" into the late twentieth century has failed to recognize how this tradition has reinscribed retrograde and oppressive attitudes toward women and domesticated animals. Greta Gaard, in "Toward New EcoMasculinities, EcoGenders, and EcoSexualities," argues that ecofeminists need to develop anti-essentialist, intersectional models of gender and sexuality to account for the variety of contexts that ecologically committed people of all genders experience and to counter capitalist heteromasculinity that is one of the central forces destroying the earth and the animals that live here.

Ecofeminist theory provides ethical guidance to challenge inequities arising along racial, gendered, and species boundaries. At a time when human violence and encroachment as well as climate change threaten to permanently alter the earth, with devastating consequences for all the animals and plants that make this planet home, the insights of ecofeminists are more important now than ever.

1892. Edith Ward's review of Henry Salt's *Animal Rights* published in *Shafts*, a British feminist newspaper.

Groundwork

Carol J. Adams and Lori Gruen

Prologue: Intersectional commitments

"Intersectionality" has become a bit of a buzzword among feminist social justice scholars and activists these days. Some have warned that moving "intersectionality" beyond the context in which it was developed as a critical race theoretical intervention into anti-discrimination law (Crenshaw 1989) risks it becoming "a flat geography," dulling rather than illuminating power relations (McKittrick 2006; Caratathis 2013). Black feminists are keenly aware of the ways in which species is racialized and race is animalized, most glaringly in the context of black women's sexuality.[1] So carefully employing intersectionality as a method for analyzing and combatting oppressive structures and practices beyond discrimination law is crucial to broader social change (Deckha 2008). Analyzing mutually reinforcing logics of domination and drawing connections between practical implications of power relations has been a core project of ecofeminism, even before the word "ecofeminism" was coined.

In 1892, for example, when a working-class, feminist newspaper reviewed Henry Salt's book, *Animal Rights*, for *Shafts*, the writer, Edith Ward argued that "the case of the animal is the case of the woman." She explained that the "similitude of position between women and the lower animals, although vastly

1906. A statue to the Brown Dog unveiled in London, leading to the protests that involved suffragettes, labor unions, and anti-vivisectionists working together.

different in degree, should insure from the former the most unflinching and powerful support to all movements for the amelioration of the conditions of animal existence. What, for example, could be more calculated to produce brutal wife-beaters than long practice of savage cruelty towards the other animals? And what, on the other hand, more likely to impress mankind with the necessity of justice for women than the awakening of the idea that justice was the right of even an ox or a sheep?" (Ward 1892, 41).

Hilda Kean in "The Smooth 'Cool Men of Science'" shows how animal issues, especially anti-vivisection, were linked to the rise of feminism and socialism from the late nineteenth century. In general, Kean argues, animal issues were picked up by progressive rather than reactionary forces. Frances Power Cobbe, a forceful activist who protested vivisection and wife-beating, recognized "the connections between vivisection, pornography, and the condition of women" (Kean, 129). She participated in protests against vivisection, and founded the Victoria Street Society for the Protection of Animals Liable to Vivisection. Kean also shows how "The anti-vivisection campaigns themselves recalled both earlier feminist 'exposure' tactics, as used by Josephine Butler in her crusade against the Contagious Diseases Act, and the 'new' tactics of public demonstrations" (26).

Two centuries earlier, stirrings of connections were articulated when the Leveller Richard Overton made a case for animals going to heaven as part of his argument for political levelling in human society—both forms of equality (Fudge 2000). In her writings, Margaret Cavendish, the Duchess of Newcastle directly challenged Descartes about his view of animals as machines. With her *Philosophical Letters* and her *Poems and Fancies*, she was one of the first to articulate the idea of equality for animals. Though he held conservative views about women, abolitionist William Wilberforce also worked for prison reform, to improve the working conditions for chimney-sweeps and textile workers, and in 1824 became one of the founders of the world's first animal welfare organization, the Society for the Prevention of Cruelty to Animals. Of the activists of the 1790s, James Turner reports: "Radical politics and other unorthodox notions went hand-in-glove with their vegetarianism" (19).

Remarkably, in the early twentieth century, anti-vivisection riots in London united trade unionists, feminists, and animal advocates. Groups that had often competed for public attention were linked in protest against the vivisection of a dog. Through their experience of being subjected to torturous forced feeding during hunger strikes in prison, English suffragettes found themselves identifying with vivisected animals. Coral Lansbury, in her important book, *The Old Brown Dog*, explained: "Women were to be the strength of the antivivisection movement, and every flogged and beaten horse, every dog or cat strapped down for the vivisector's knife reminded them of their own condition." And not just suffragists—workers, too, found themselves identifying with animals as victims. Activists and writers both saw "workers and animals sharing the same fate" (Lansbury, 82).

Isabella Ford's pamphlet "Women and Socialism" was published a year after the Brown Dog riots. Hilda Kean explains, it

> is perhaps best known for the links she makes between class and sex oppression. Yet the connections she draws between the experiences of women and domestic animals are also perceptive. Isabella Ford laments the effects of industrialisation that has led to a misunderstanding of nature. She evokes the experience of non-human animals to illuminate the experience of women: "In order to obtain a race of docile, brainless creatures, whose flesh and skins we can use with impunity, we have for ages past exterminated all those who showed signs of too much insubordination and independence of mind" (Kean, 29).

The same year of the Brown Dog riots in London, across the Atlantic at the National American Woman Suffrage Association (NAWSA) annual meeting, a vegetarian milliner requested that the Treasurer of the NAWSA stop wearing bird feathers in her hat. This suffragist declared: "Nothing would persuade me to eat a chicken, or to connive at the horror of trapping innocent animals for fur" (cited in Adams 1990, 2010, 224).

The commitment to non-violence guided the civil disobedience that first the British suffragettes engaged in, and then the American suffragists. It was their civil disobedience that brought about their imprisonment and subsequent force-feeding. Gandhi, living in England at the time, was inspired to take the tactics of the suffragettes back to India to employ in his challenge to British colonial power.

Feminists have a long history of acting on shared commitments with other social justice and non-violent movements. In this essay, we highlight the intersectional commitments of ecofeminist theory and practice through a contextual analysis of events, writing, and reflections of activists and scholars. Partly we do this to pay tribute to the historical insights of these thinkers and activists; partly we do it to provide a record of how we view the development of ecofeminism as we move theory and practice forward.

Activism in the 1970s and 1980s

The Civil Rights and anti-Vietnam war movements in the 1960s paved the way for the second wave feminist movement and by the early 1970s,

1907. Isabella Ford's pamphlet "Women and Socialism" appears in a publication by the International Labour Party.

feminists were establishing theoretical and activist connections with animal issues. Tracing human mistreatment of other-than-human animals to patriarchy, feminists in the 1970s were identifying not only the ways that women and animals were thought to be less valuable by mainstream patriarchal culture, but also were arguing that emerging animal activism required a feminist analysis.

In the early 1970s, Connie Salamone, a part of the Vegetarian Activist Collective, began to write about the connections between the experience of the other animals and of women. By 1974, Salamone had travelled around the United States and to Europe, "to bring her plea of including an interspecies solidarity into the emerging feminist manifestoes" (McAllister, 364). She started "Vegetarian-Feminists" which urged feminists to resist the exploitation of animals and to stop eating meat.

During 1975 and 1976, Carol Adams interviewed more than forty feminists who were also vegetarians in order to uncover the reasons they had adopted a vegetarian diet. Many of them articulated an ecofeminist perspective that located animals within their analysis. One said, "Animals and the earth and women have all been objectified and treated in the same way." Another explained that she was "beginning to bond with the earth as

Figure 1.1 Connie Salamone (far left) and other members of the Vegetarian Activist Collective in the early 1970s protesting against fur in front of a mid-town Manhattan. Photograph © by Karen D. Messer. Used by permission.

a sister and with animals as subjects not to be objectified." A third reported, "Feminists realize what it's like to be exploited. Women as sex objects, animals as food. Women turned into patriarchal mothers, cows turned into milk machines. It's the same thing" (Adams 1991: 89–91).

Some feminists in the early seventies were influenced by their anti-war, non-violence activism to become vegetarians and began thinking about the relationship between the oppression of women and people of color both in the US and overseas, as well as animals and the earth. The 1971 publication of Frances Moore Lappé's *Diet for a Small Planet* was also influential. As one of the feminists Adams interviewed noted, "by eating meat you are exploiting the earth and to be a feminist means not to accept the ethics of exploitation."

In 1976, feminists organized a Women and Spirituality Conference in Boston; this was the first feminist conference that provided only vegetarian food and featured a workshop on feminism and vegetarianism, although not without some angry responses and letters to the Washington-based feminist newspaper, *Off Our Backs*. Feminists were also inspired by the emergence of ecofeminism and writings such as Rosemary Ruether's *New Woman, New Earth: Sexist Ideologies and Human Liberation*, in which she argues that patriarchal civilization is built upon the historical emergence of a masculine ego consciousness that arose in opposition to nature, which was seen as feminine.

Ecofeminism and the Challenges of Globalization
Rosemary Radford Ruether

Among ecofeminists the connection between the domination of women and the domination of nature is generally made on two interconnected levels, the cultural-symbolic level and the socio-economic level. Among Western ecofeminists the connection has usually been made first on the cultural-symbolic level. One charts the way in which patriarchal cultures of the West have defined women as being "closer" to nature, as being on the nature side of a nature-culture hierarchy. This is shown in the way in which women have been identified with body, with earth, with sexuality, with flesh in its mortality, in its presumed weakness and sin-proneness, *vis à vis* a construction of masculinity identified with spirit, mind and sovereign power.

A second level of ecofeminist analysis goes beneath the cultural-symbolic level and explores the socio-economic underpinnings of this ideology of women's similarity to non-human nature. How has the domination of women's bodies and women's work by ruling class men been interconnected concretely with the exploitation of land, of water, or animals? How have women as a gender group been colonized by patriarchy as a legal, economic, social and political system? How does this colonization of women's bodies and work function as the invisible and unrecognized substructure for the extraction of natural resources for the enrichment of the male ruling class? How does the positioning of women as the caretakers of children, and of small animals, the gatherers of plants, the weavers, the cooks, the cleaners, the waste managers for men in the family, function to both inferiorize this work and to identify women with a non-human world that is likewise inferiorized?

Such ecofeminist analysis reveals the way that the cultural and symbolic patterns by which both women and nature are inferiorized and identified with each other function as the ideological superstructure by which the economic-social-legal domination of women, land and animals is justified and made to appear "natural" and inevitable within the total patriarchal cosmovision. Elite males, in different ways in different cultures, create hierarchies over subjugated humans and non-humans, men over women, whites over Blacks, ruling class over slaves, serfs and workers.

These structures of domination between humans mediate the domination of elite males over non-human nature. Women were traditionally subjugated to confine them to the labor of reproduction, childcare, and productive work that turns the raw materials of nature into consumer goods, while being denied access to education, culture, control of property and political power of the ruling group, whose roles are then identified with "human" transcendence over nature.

What this means is that women's inferiorization to men is modeled after the inferiorization of non-human nature to "man," and vice versa. The term "man" itself is then understood to be an androcentric false generic which actually means the elite male as the normative human being, with women, slaves and peoples of other races and cultures seen as lesser humans or subhumans, identified as standing between mind and body, human and animal. This is the way Aristotle, for example, in his *Politics*, understands the relation between elite Greek males, and women, slaves and barbarians, who for him are "natural slaves," much like animals, tools or land, ultimately like the body in relation to the mind, in his dualistic philosophical world view.

This interconnection between the subjugation of women and that of subjugated races and classes means that ecofeminism cannot treat women as a univocal category. Women are a gender group within every class and race. That means they share in the privileges or disprivileges of their class and race, while also being inferiorized as women in relation to men within their class and race. But this disprivileging as women, *vis à vis* the men of their class and race, obviously takes very different forms across classes and races. The women of the 19th century Boston Brahmins were expected to supervise female and sometimes male servants, who might be Irish or Black, while not allowed to pursue higher education at Yale or Harvard or to aspire to a career in business, politics or the church.

Women servants or slaves experienced much more oppressive lives in every way. Since women of the elite were often their most immediate oppressors, it is hard for such Black or working class women to see themselves as sharing a common oppression with white elite women. It takes a bit of perspective to recognize that all these women, as well as male servants and workers, are part of one system designed to place different groups in different roles across class and race for the benefit of one master group, elite white males.

People today, unaccustomed to such ecofeminist analysis, may be inclined to dismiss it as either exaggerated or passé, as they look to the way in which women of elite classes and races have won their way into something like the privileges of their brothers. In the United States these daughters and sisters of elite men, and some of not such elite men, now go to Harvard and Yale, or the University of California, or Cal State and aspire to careers in business, the church and politics, although they are still a small minority in the upper reaches of these careers.

Far from making ecofeminism irrelevant, these efforts to show the complexity of gender within class and race reveal why ecofeminism must interconnect with the movements of environmental racism and ecojustice and situate itself in a global context. By environmental racism, I mean those movements among African Americans and Indigenous peoples (mostly spearheaded by women of these groups) that are struggling against toxic dumping and environmental pollution that is concentrated particularly in the areas where poor people of color live. Global ecofeminism shows how this pattern of impoverishment of nature and of emiserated humans are interconnected in a worldwide economic system skewed to the benefit of the rich beneficiaries of the market economy.

Whenever gender is analyzed across class and race world wide the reality that women are still the poorest of the poor becomes starkly evident. An essay on women in relation to world population in the 2002 State of the World Report makes this clear. Two thirds of the world's 876 million illiterate people are female. In 22 African nations and 9 Asian nations school enrollment for girls is less than 80% that of boys. Only 52% of girls stay in school past the fourth grade in these countries. Only about 4 women per 1000 attend high school, much less college.

Even in the United States, where 18% of households are headed by women, these households account for a third of the children living in poverty. Women throughout the world earn significantly less than men, on average between two-thirds and three-fourths, and women account for only 5% of the senior staff of the 500 large corporations. Only in 9 nations are women 30% or more of the members of parliament, six of them in Africa, (Rwanda, Burundi, Mozambique, South Africa, Tanzania and Uganda) while in the Americas women hold only 15% of parliamentary seats and only 4% in Arab states. In many states women are still legally under the guardianship of their husbands or fathers and have no right to manage property.

Physical abuse shadows women from before birth. Sex-selected abortions, female infanticide, malnutrition and abuse of female children are common in many nations. In India dowry murders, the killing of wives in order to seek dowries from second wives, continue to happen despite 25 years of efforts to expose this practice. Incest, female genital cutting, denial of medical care, early marriage, forced prostitution and forced labor hang over women's heads world wide. Girls and women are more likely to be sold into slavery than males, and some 130 million women world wide have experienced the cutting of their genitals, a practice that continues at the rate of about 2 million a year.

Those concerned with both population and environment have recognized that the single factor most likely to both check population expansion and also improve the health and welfare of children in families and care for the local environment is the promotion of the equality of women with men. Americans are likely to assume that women and children share in affluence or poverty pretty much on the same level with men of their family. But, in fact, studies continually show that men tend to use the majority of their own assets for themselves, not for women and children of their own families. Women, by contrast, devote the great majority of the fruits of their own labor, whether in cash or in subsistence labor, to feeding, clothing and educating their children. Women also do much of the subsistence labor that protects and renews local environments. Thus increasing women's share in education, income and power is a major factor in improving the health, welfare and education of children.

We must be clear that promoting women's equality is not a matter of separating women from men or children, as is presumed in the United States. Rather its purpose is to convert the relation of women and men to a greater partnership and sharing of the care of households and children, tasks which presently fall disproportionately on the shoulders of women. Feminism has been misconstrued, sometimes by feminists themselves, primarily in terms of women gaining rights and access to the same alienated and dominating roles as men, sharing the same fiction of autonomous individualism.

We need to continually insist that feminism is about the converting of patterns of patriarchal domination for women and men into a new relationship of mutuality. Not only is this not anti-family, but in fact families and particularly children are the first beneficiaries of such restoration of men to caring relations with women and children. Likewise ecofeminism must not be seen as making women the primary caretakers for the local environment, but bringing men into the work of care for the household and earth, now being borne disproportionately by women. Why is it that it is predominantly women, rather than men and women together, who have led the struggles against toxic environments and who do the bulk of the recycling of wastes? It is essential that ecofeminism be integrated with socioeconomic analysis and struggle against the structures of impoverishment of poor women and men and the earth.

For example, African women are developing concerns for ecology in the context of their native traditions. In my book *Women Healing Earth: Third World Women on Ecology, Feminism and Religion* I discuss the views of six African women on ecology. Denise Ackerman a white Christian feminist theologian and Tahira Joyner a Muslim scholar of religious studies, both from Capetown, South Africa, compare the involvement of their religions in ecological justice in their country. Although Christians and Muslims played key roles in the struggle against apartheid, environmental destruction and its relation to poverty and racism has yet to be acknowledged adequately. But there are a number of NGOs that are involved in this issue. Tsepho Khumbane has organized on reforestation and water resources in the Transvaal and Nam Rice has been a leader on protection of dolphins and other sea animals in the Cape coastal area.

1915. Four American vegetarian feminists travel on the Ford Peace Ship from the United States to Europe.
1915. Charlotte Perkins Gilman's *Herland* published, which depicts a feminist-vegetarian-pacifist utopia.

Another activist on ecological thought is Sara Mvududu, a Zimbabwean sociologist who works on environmental protection and gender issues with Women and Law in the Southern Africa Research project in Harare. Both traditional Shona patriarchy and the racist patriarchy brought by British colonialism marginalized rural African women. But she argues that both their traditional work as cultivators of the land and new modern rights can empower rural African women in Zimbabwe to become proactive agents of environmental sustainability. Isabel Phiri, a lecturer in theology at the University of Zomba, Malawi, examines the Chisumphi cult in her country where rural women were in charge of practices of rainmaking and cultivation of the earth. She discusses ways in which the implicit environmental ethic of this cult can be developed to respond to the new situation of ecological crisis. Tumani Nyajeka, from Zimbabwe and Theresia Hinga from Kenya also write to show how indigenous religious traditions that value human relations to nature and care of the land can be resources for contemporary concern with ecology in their African contexts.

There is emerging world wide radical revisions of the way in which traditional patriarchal religions have been understood to reinforce the domination of men over women, "man" (male elites) over nature. Feminist Christians, Muslims, Jews and Buddhists, among others, are rethinking their inherited religious traditions to mandate more egalitarian and mutual relations between men and women and humans and nature. In the Christian tradition this takes the form of several key revisions of traditional Christian theological symbolism.

One may start with how humans have envisioned themselves, particularly how they have seen the relationship of mind to body, human intelligence to nature. Mind or consciousness does not originate in the stars, as Plato thought, nor is it infused into bodies by a transcendent God outside the universe. Rather human consciousness is an intensification of interactive awareness that exists to some degree on every level of reality, from sub-atomic physics to organic molecules to photosynthesizing plants to increasingly aware and communicating animals. We might think of our particular human gift of symbol making consciousness as the place where earth becomes aware of itself in a new self-reflective way. This does not separate us from other species, but calls us to be the place of celebrating the cosmic process. We also need to use our intelligence to learn to harmonize our human needs with those of the rest of the earth community.

We need to shape new households that model just and sustainable relations between humans, starting with men and women, and with the natural world. These household communities need to be attractive models of sustainable technology, harmonized social relations and celebrative culture. We need to shape these households, not to withdraw into them, but as bases for networking an alternative economy and society to reshape the larger global systems into a sustainable earth community. This struggle to reshape the death systems of our world cannot flow only from anger, fear and guilt. It must be more deeply grounded in joy in the goodness of life and gratitude for its gracious vitality. To create glimpses of health and joy is our basic task as ecofeminists.

Rosemary Radford Ruether, a feminist theologian, has been exploring intersectional oppressions since the 1970s, when her book, *New Woman, New Earth: Sexist Ideologies and Human Liberation* appeared. She is the author of *Gaia and God: An Ecofeminist Theology of Earth Healing* (1994); *Sexism and God-Talk: Toward a Feminist Theology* (1983); *America, Amerikkka: Elect Nation & Imperial Violence* (2007), and other books, most recently, *My Quest for Hope and Meaning.*

As feminists were developing what we now call "ecofeminism" in the 1970s, the contemporary animal rights movement was beginning to take shape. By the early 1980s as more and more animal advocacy groups emerged it was becoming clear that the budding movement, though rhetorically attuned to sexism and racism, was not as inclusive as it should have been. Throughout the movement, white men assumed leadership roles, while mostly white women were often relegated to sex-stereotyped roles. The visible spokespeople, theoreticians, and writers were overwhelmingly white and male. Organizations operated with hierarchies, in which, as Marti Kheel put it, a "small, elite group makes all the major decisions which the rank and file obediently carry out" ("A Feminist View of Mobilization," 1984, 2).

Ecofeminists concerned with other animals began arguing that animals are individuals, with feelings, needs, and the capacity to love and to suffer. They were critical of the way that major animal rights events (like the April 24, 1983, Mobilization for Animals during which 15,000 people assembled at four major locations targeting the Primate Research Centers) focused on celebrity speakers. Reflecting on the protest at Davis, where Bob Barker spoke at the rally (one woman refused to go "because she did not want to hear Bob Barker, a man whom she stated 'had been exploiting women on television for years'"), these events led Kheel to write:

> But is there not something strange about a movement that feels compelled to establish its credentials through association with the "stars"? Haven't feminists been involved in enough movements that produced 'stars' or were run by leaders? Isn't this, after all, one of the aspects of patriarchy that feminists are fighting against? (1984, 2)

She juxtaposed this rally with the Women's Pentagon Action of 1980 where there were no celebrities and women were able to learn from each other. Women used civil disobedience to block the doorways to the Pentagon and in one protest weaved yarn across one of the main entrances. "In short, women's collective energy was unleashed in many creative ways." Kheel concluded that "Women must be aware of the problems inherent in working in a movement [the animal rights movement in this case] dominated by men."

Sally Gearhart, whose utopian novel *The Wanderground*, published in 1978, had immediately become a lesbian-feminist classic, began to speak on behalf of animals shortly thereafter. She spoke on interconnected oppressions at the 1981 rally on World Day for Laboratory Animals at Letterman Army Institute of Research, San Francisco: "I'm here because I now see that there are fundamental connections between women's rights and the rights of nonhuman animals, and I want to talk about those connections, about the dehumanization that I feel is going on in all of us, and about my own change from being just a lesbian and a feminist activist, to being an animal activist as well."

Circa 1915. Feminist, vegetarian, pacifist Charlotte Despard offered vegetarian meals at the cheap meals service she offered on her property.

Feminists for Animal Rights
Bayta Bauman

Feminists for Animal Rights (FAR) had a dual purpose: to bring a feminist perspective to the animal rights movement, exposing sexism in the movement, and to imbue an animal-caring consciousness in the feminist movement. On the one hand, the feminist movement knew all about patriarchy but made no connections between the ways animals are treated and the patriarchal values it worked, indeed existed, to replace. On the other hand, the animal rights movement knew all about how horribly animals are treated but knew little or nothing about patriarchy—the rule of the fathers which existed from ancient times right up to the present—and its values and perceptions that are responsible for such mistreatment. To the animal rights movement, the understanding FAR brought was that the perception and treatment of animals was a patriarchal equivalent to its perception and treatment of women. Or, it could be said, the other way around, that the treatment of women, in patriarchy has been similar to the perception and treatment of animals.

It is hard for me to admit, considering myself a radical feminist, that work in the feminist movement was much more difficult than work in the animal rights movement. There were men in the movement who understood, who really got it. One powerful message that FAR brought to feminists, exactly the people who should most understand it, was that, in addition to eating meat, eating dairy and eggs meant participation in industries that manipulate and exploit female reproductive processes.

I feel fortunate to have participated in what is perhaps the two most important movements of our times—the feminist movement and the animal advocacy movement. Some may argue this choice, and point out other movements that they consider as important or more important, but the thing about both feminism and animal advocacy is that they are both "primary" movements from which ideas flow to other causes and movements. Both deal with basics. There can be no peace, racial equality, civil liberties, environmental justice without resolving issues of exploitation and abuse of those perceived as "other."

Batya Bauman, a lesbian-feminist activist, originally conceived of, and is one of the founders of, the Jewish feminist magazine, *Lilith* (1976), and directed the activities of Feminists for Animal Rights, including editing its Newsletter, from the late 1980s.

Around the time FAR started, other feminist groups interested in animals began forming: *World Women for Animal Rights*, founded by Connie Salamone, the *Canadian Feminists for Animal Welfare*, the *Australian Feminists for Animal Rights*, and the *British Women's Ecology Group*, which had a special emphasis on "understanding the present human predicament and mass animal suffering (vivisection and factory farming, etc.)" from the perspective of the "systematic crushing of the Feminine Principle by patriarchal power."

Not all feminists were in agreement about every issue; indeed there were many areas of conflict and tension. One of the intense debates that emerged in the 1980s came to be known as "the sex wars." Some feminists and ecofeminists were fundamentally opposed to pornography and sex work, believing it perpetuated the eroticization of subordination and led to violence against women. Others believed that women's sexuality had been coopted by dominant forces for too long and were interested in embracing the "pleasures and dangers" of non-normative sexual expression. Ecofeminists divided on these issues as well, and the small study group that led to the formation of Feminists for Animal Rights (FAR) viewed sex work as oppressive.

Some feminist writers have argued that man's control over women, animals, and nature is definitional to patriarchy. Marilyn French's *Beyond Power* argues, "patriarchy is an ideology founded on the assumption that man is distinct from the animal and superior to it. The basis for this superiority is man's contact with a higher power/knowledge called god, reason, or control. The reason for man's existence is to shed all animal residue and realize fully his 'divine' nature, the part that *seems* unlike any part owned by animals— mind, spirit, or control" (341). Elizabeth Fisher's *Woman's Creation: Sexual Evolution and the Shaping of Society* argues that "the sexual subjugation of women, as it is practiced in all the known civilizations of the world, was modeled after the domestication of animals ... Animals ... may have been the earliest form of private property on any considerable scale, making animal domestication the pivot also in the development of class differences."

Despite these observations, the environmental movement, as well as those developing its theoretical foundations, attempted to elevate nature at the expense of animals, women, and people of color. In 1985, Kheel published a stunning critique in *Environmental Ethics:* "The Liberation of Nature: A Circular Affair." Kheel was responding to environmentalists in general and to J. Baird Callicott's statement that "Environmental ethics sets a very low priority on domestic animals as they very frequently contribute to the erosion of the integrity, stability, and beauty of the biotic communities into which they have been insinuated." She challenged the concept of holism that viewed "the 'whole' as composed of discrete individual beings connected by static relationships that rational analysis can comprehend and control." Kheel recommended that writers in "environmental ethics might spend less time formulating universal laws and dividing lines, and spend more time using reason to show the limitations of its own thought."

Andrée Collard's *Rape of the Wild: Man's Violence against Animals and the Earth* (1988) focused on animal experimentation and hunting. Collard saw animals as a window to the death-oriented values of patriarchal society, both because she was deeply concerned for animals' wellbeing "and partly because man's treatment of them exposes those values in the crudest, most undisguised form" (2).

These critiques led to increased scrutiny of the rationalist bias in Tom Regan's "animal rights" and Peter Singer's "animal liberation" theories, on the one hand, and the hierarchical bias in the holistic ethics of Aldo Leopold and J. Baird Callicott. Increasingly ecofeminist theorists were interested in developing an understanding of the role of affect, that fused reason and

1917. Susan Glaspell's short story "A Jury of Her Peers" is published, in which women who accompany their law officer husbands to the scene of a crime recognize that the death of a bird was central to what happened.

emotion, and was rooted in a personal sense of loving, caring connection with all life-forms.

By the end of the 1980s, a kind of activism that directly connected to affect and the ethics of care appeared: animal sanctuary. Sanctuaries for farmed animals put loving, caring connection into practice and provided opportunities for the public to get to know the animals who were usually thought of as meat. Animals rescued from slaughterhouses, factory farms, and auction houses, were given opportunities for new, unshackled lives. In 1985, Karen Davis discovered a flock of chickens in a shed on a piece of property she rented in Maryland. Six weeks later, only one was left, Viva; the others had been taken to slaughter. Viva became the catalyst for the creation of a sanctuary, and was crucial in Davis's decision to found *United Poultry Concerns,* a nonprofit organization dedicated to promoting the compassionate and respectful treatment of chickens and other domestic fowl. Davis's engagement resulted in her critique of environmentalism's masculine mystique in a 1989 article in *Between the Species.*

In 1986, Gene Baur and Lorri Bauston founded Farm Sanctuary outside of Watkins Glen, NY. The Sanctuary movement recognized that creating opportunities for people to experience interactions with the animals who usually were seen as "dinner" was an important way of doing vegan outreach. It also served to make what Adams calls the "absent referent"—the animal who is consumed and available only as a slaughtered body—present.

Figure 1.2 UPC sanctuary roosters Ivan (center), Lorenzo (left), and Benjamin (right) with their blissfully dustbathing hens on May 22, 2013. Photograph © by Richard Cundari. Used by permission.

Providing sanctuary to poultry
Karen Davis

I recently adopted a young rooster I named Ivan into our sanctuary in Machipongo, Virginia. I love watching the interactions between newly introduced birds and those who are already well established and comfortable in their surroundings and in their social relationships with one another and with me. There is a dance of life and perpetual drama going on amongst our rescued birds, all of whom have in one way or another been "scathed" by people, but who had the luck to find permanent refuge at United Poultry Concerns.

I defy any observer to say that these birds do not fully enjoy the ground under their feet, the sunlight which they follow and sunbathe under on winter days and every day, the small breezes that ruffle their feathers when they are sitting quietly together under the trees in the middle of the afternoon. I dare anyone to say they are not fully alert and conscious of themselves, their surroundings, and each other. Let a hawk soar overhead, or a fox sneak close to the fences, and you will see and hear in an instant whether these birds have more in common with tables and chairs than they have with their wild relatives. So how does Ivan fit himself into this charmed community of avian life? During the first few weeks of his arrival, the top roosters, Rawley and Benjamin, intermittently run him off whenever he tries to get too close to the inner circle. Ivan, meanwhile, who understands in his bones what is going on, is watching, waiting, and making his moves, and they are doing likewise, everyone being fully involved in the shifting patterns and rhythms of their social adjustment.

In the evening, Ivan watches, and just as attentively, he listens outside in the yard, for Benjamin to finally fly up to his perch on top of the heavy wooden door, and for Rawley to ascend to his cabinet post in the Big House from which he will not fly back down once he is settled up there with his hens on either side of him. Cautiously, Ivan assesses the situation, and when he perceives that the coast is clear, he enters the enclosure and makes his way to a chicken ladder rung where, I can confidently say, he hopes that at least one hen will be sitting for him to sit next to during the night.

Karen Davis is the founder of United Poultry Concerns which works on behalf of domestic fowl—including chickens, turkeys and ducks. She runs a sanctuary and is the author of *More Than a Meal: The Turkey in History, Myth, Ritual, and Reality* (2001), *The Holocaust and the Henmaid's Tale: A Case for Comparing Atrocities* (2005), and *Prisoned Chickens, Poisoned Eggs: An Inside Look at the Modern Poultry Industry* (2009).

Making the connections: The 1990s

If anything stands as a touchstone for a moment when feminism and animal activism came together, it is the 1990 March on Washington. Fifty thousand people marched. Seventy-five percent of the marchers were women, six out of 20 speakers were women; three of them were performers including Laura Nyro and Grace Slick. Feminists for Animal Rights commissioned a lavender banner that cost $350 and chartered a bus for 48 women to travel from New York City. Georgia Lesbian Ecofeminists for Animal Rights arrived from Atlanta organized by Denise Messina. A woman watching the parade surprised Denise Messina by rushing up to her and hugging and kissing her, explaining she was so moved to see the banner and the feminist

1938. Virginia Woolf's anti-war *Three Guineas* includes this aside: "Scarcely a human being in the course of history has fallen to a woman's rifle; the vast majority of birds and beasts have been killed by you, not by us."

Figure 1.3 The Feminists for Animal Liberation banner for the 1990 March on Washington. In the center are Batya Bauman, Carol Adams, and Marti Kheel. Photograph © by Bruce A. Buchanan. Used by permission.

presence. Through the activism of Batya Bauman, Carol Adams was invited to be one of the speakers. Ingrid Newkirk was another speaker.

Several things were happening in the early 1990s—as the publications listed in our timeline indicate, ecofeminism was engaging in a robust dialogue and emerging as an important philosophical and theoretical approach. Groups like EVE (Ecofeminist Visions Emerging) formed a "guerilla grafitti" group that transformed anti-woman and anti-animal advertising around New York City with their magic markers.

In 1989, the Ecofeminist Task Force was founded to be a part of the National Women's Studies Association. Four members of the task force— Marti Kheel, Batya Bauman, Stephanie Lahar, Greta Gaard—drew up a Resolution for Dietary Nonviolence at NWSA conferences. Gaard describes the Resolution:

> For their budget and total cost of implementation, we wrote "estimate unknown; total cost of non-implementation includes the health of humans, the lives of non-humans, and the well-being of the planet." Under "other costs" we wrote "total food costs will be reduced," referring to a two-page list of "whereas" facts that describe the human health costs (breast and ovarian cancer risks, bone loss and osteoporosis, kidney stones, obesity, diabetes, heart disease, asthma, and more), the ecological costs

(deforestation, water use, species loss, soil erosion, excrement pollution), the costs of overconsumption in producing real material hunger for humans, and the costs to the "food" animals themselves. We concluded with the recommendation that "the Coordinating Council make a strong statement of feminist nonviolence, and make NWSA a model of environmental and humane behavior by adopting a policy that no animal products—including the flesh of cows, pigs, chickens, and fish, as well as all dairy and eggs—be served at the 1991 conference, or at any future conferences."

The recommendations were clearly for a vegan diet, but in the early 1990s, "vegetarianism" remained the prevalent descriptive term. Gaard explains what happens at the 1990 NWSA Conference:

Leading up to the conference, strong allegations of racism and classism within the national office were discussed across caucuses and members. Concerns simmered and finally erupted at the close of the conference: when the NWSA Coordinating Committee insisted on following Robert's Rules of Order and failed to respond to the Women of Color Caucus concerns by late afternoon, the Women of Color Caucus walked out of the meeting and the organization, and the Lesbian Caucus walked out in solidarity. All items on the delegate assembly's agenda were put aside, including the Ecofeminist Task Force's resolution, and in a separate meeting after the conference—when no members could attend, advocate, or explain their proposed resolutions—the resolution was voted down. Leaders at NWSA decided not to have another national conference in 1991, and the organization was eventually re-organized to ensure that women of color were placed in primary leadership positions.

In the early 1990s, writing that drew connections between various forms of oppression was blossoming: Gaard published her collection *Ecofeminism: Women, Animals, Nature* (1993). Susanne Kappeler's "Speciesism, Racism, Nationalism… or the Power of Subjectivity" (1995) examined how a hierarchical culture structured on race, gender, and species operates. In *An Unnatural Order: Why We Are Destroying the Planet and Each Other* (1997), Jim Mason argued that all human oppression originates in the domestication and subsequent oppression of animals. Mason coined the term *misothery* (from the Greek words for *hatred/contempt* and *animal*) to denote speciesist derogation of animals, which he views as parallel to *misogyny*, hatred and contempt for women (163–4). The theory that the oppression of women, animals, and other subjugated groups are interrelated has proven to have practical implications.

1944. Dorothy Morgan coins the term "vegan" (the beginning and ending of the word *vegetarian*). She and Donald Watson marry, help to found the Vegan Society, and promote veganism as a world view and a word.

On Recognizing Interconnections
Kim Stallwood

Up until the early 1990s I was an insufferable animal rights activist tanked up on self-righteous indignation about how animals were treated. If this meant being liberal with the truth, so be it. If it also meant naming and shaming those who exploit animals, so be it. If it meant trampling on the interests of others and behaving disrespectfully toward them, so be it. Animals were suffering now. Something had to be done. Someone had to do it. And if this meant me, so be it. And if this meant you were insulted and shocked, so what?

I went to work for PETA in 1987 because I was greatly impressed with their two-part strategy of simultaneously presenting the problem and the solution. The problem of institutionalized animal exploitation was graphically displayed in their innovative undercover investigations. The solution was offered in their attractive vegan public education lifestyle campaigns.

Along with the co-founders and staff, I helped to develop and implement PETA's strategy of actions designed to attract media attention. We openly called ourselves 'media whores.' Our objective was media coverage. PETA did not have the same capacity as the media to publicize animal cruelty and exploitation. It was our job to get the media to do it. This inexorably led us to making decisions about publicity stunts whose sole purpose was to get media coverage regardless of whether it was positive or negative. And, in some cases, arguably incompatible with animal rights philosophy. All coverage was worthwhile media, even if it meant framing PETA and animal rights in a negative light and even if it meant our ads were banned because they were too controversial. In fact, we fell in love with controversy. Outrage became our friend. And we were determined to push the boundaries of reason and decency to its limits. It was all for the animals.

I went along with this journey willingly; however, I began to worry about the direction that the organisation was taking and my relationship with its co-founders began to sour. As with many others who have worked at PETA, I left.

So, the early 90s began a period of introspection in which I questioned the assumptions I had made and lived by for many years as an animal rights activist. Among my assumptions was a commitment to understanding the philosophical arguments with respect to our relations with animals. I found myself unknowingly transitioning from animal activism to animal advocacy, I became intrigued with what was I was learning from my discovery of ecofeminism.

Two developments stand out. The first is the publication of Carol's *Sexual Politics of Meat* in 1990. The second is watching Marti and Carol present a Feminists for Animal Rights slide show. They awakened in me new and intriguing ways to think about animal rights and animal advocacy. I liked how ecofeminism presented animal exploitation within a progressive context alongside other social justice issues. I found it exciting to see written and visual analysis being made of ideas about masculinity and masculine behaviour. As a gay man who was intuitively uncomfortable with sexism but little understood the theories of feminism, they resonated well and deeply with me.

Now ecofeminism helps me to understand my work for animal rights as a practice. Embedded into my practice are ecofeminism's understanding and values.

Kim Stallwood has been an animal activist for over 35 years. He was the editor of *The Animals' Agenda* magazine (1993–2002); edited two anthologies: *Speaking Out for Animals* (2001) and *A Primer on Animal Rights* (2002), and is the author of *Growl: Life Lessons, Hard Truths, and Bold Strategies from an Animal Advocate* (2014).

Feminists were working to help victims of domestic violence find shelter with their companion animals. Aware of the challenges for victims of domestic violence who did not want to leave their companion animals, FAR launched a "Companion Animal Rescue Effort" (CARE) program and offered support to other groups on how to start programs ("Feminists for Animal Rights," 1994).

Adams's "Woman-Battering and Harm to Animals" (1995) was one of the earliest essays providing a feminist theoretical framework for the injury and killing of companion animals by batterers. It was used for many years as the source for the Humane Society of the United States's website on domestic violence and animals. After identifying the ways in which batterers harm animals and the effects upon the battered women, Adams explored the implications for feminist theory: "Gender is an unequal distribution of power; interconnected forms of violence result from and continue this inequality. In a patriarchy, animal victims, too, have become feminized. A hierarchy in which men have power over women and humans have power over animals, is actually more appropriately understood as a hierarchy in which men have power over women, (feminized) men, *and* (feminized) animals" (1995a, 80).

While more writings making connections continued to appear, some theoretical tensions between feminists and ecofeminists also emerged. For example, *Signs*, a leading feminist theory journal, rejected a review essay by Gaard and Lori Gruen though *Signs* had commissioned it; the editors at *Signs* wrote: "ecofeminism seems to be concerned with everything in the world … [as a result] feminism itself seems almost to get erased in the process." The article was later published in *Society and Nature* 1993 and has subsequently been anthologized.

Just as feminists came down on different sides of the "sex wars," there were also tensions about the complexities of gender expression. While ecofeminism was under criticism for appearing essentialist, it was also being criticized for being trans-exclusive. Feminists for Animal Rights contributed to this perception by having a woman-born-woman membership rule, much like the Michigan Women's Music Festival. This was the source of debate both within FAR and for those who wished to start FAR groups in their cities. The need for cis-women to create safe spaces devoted to their own concerns, particularly for those who have been sexually terrorized and abused, has been seen by some to be trans-exclusive. Ecofeminists have been exploring how to be more trans-inclusive without reinscribing painful, problematic binaries. As pattrice jones has recently written:

Ecofeminism would prescribe dialogue rooted in an ethos of care aimed at getting past either/or thinking. Can't we *both* work for the rights of people who choose hormones, surgeries, or other ways of identifying as the other sex *and* critique the notion that a boy who prefers flower gardening to hunting isn't "a real boy"? … Productive discussion will be more likely if everybody is in possession of the facts about the natural history of gender expression among human animals, which has been much more diverse than most feminist *or* trans activists realize. We understand that the currently dominant gender system—which both feminist and trans activists critique, albeit in different ways—is a product of the same European mania for pseudoscientific categorization that brought us the conceptions of race and species that are central to racism and speciesism. We know that taking that diversity into account can lead to new ways of thinking about both gender and sexuality.[2]

Feminists and ecofeminists were also split about the importance of vegetarianism. For example, in 1994, Kathryn George published a paper in *Signs* claiming that the moral demand for universal vegetarianism places women and some others in a "moral underclass" as their bodies could not be vegetarian and that when feminist-vegetarians advocate vegetarianism we are making a "virtue of our own oppression." *Signs* editors did not think to have a feminist-vegetarian scholar review its arguments before publishing it. Rather than actually addressing herself to feminist writers and their critiques of Singer and Regan, George critiqued these philosophers herself. She also argued that nutritional researchers had neglected women in their studies and that women and children did not do well on a vegan diet. In a subsequent issue, Adams (1995b), Donovan (1995), Gaard and Gruen (1995) responded, tackling many issues:

- if nutrition researchers had indeed neglected women, then there was no reliable research to claim anything about women and vegan diets, therefore George could not claim that vegan diets affected women adversely;

- in fact, current nutritional information does not support George's claim;

- she enacts the naturalistic fallacy that one constructs ethical ideals upon empirical "norms";

- ecofeminist veganism is not a universalist demand, but rather a feminist methodology that carefully contextualizes gender, race, class, sexuality, and ethnicity.

From the first feminist conference that solely served vegetarian food in the 1970s and roused criticism, to George's flawed attack on veganism in the 1990s, the debate about feminist-vegetarianism continues into the twenty-first century as the "Paleo-diet" is adopted by some feminists. Feminist Lierre Keith's *The Vegetarian Myth* attacks a vegan diet, but her research—as vegan nutritionist Virginia Messina, RD, has pointed out—is extremely faulty. For instance, she criticizes grains as a food our Paleo-ancestors would never have eaten, but this is equally true with cow's milk, which Keith recommends.[3]

Top 10 Ecofeminist Actions to Take This Summer [1996]
From the FAR Newsletter

10. Donate a copy of *The Sexual Politics of Meat* or another personal ecofeminist favorite to your local library.

9. Wear a vegetarian t-shirt to a company picnic and provide a generous amount of vegan food.

8. Teach a class in vegan cooking at your local food co-op.

7. Keep a supply of fresh water in your yard for thirsty neighborhood critters.

6. Plant an organic garden.

5. Set up a Premarin table at your local grocery store and take Premarin brochures to your gym.

4. Create a library display on ecofeminism or arrange a showing of *Ecofeminism Now!*

3. Get a tattoo of the FAR logo (somewhere you can see it!).

2. Using magic markers, correct misogynist and speciesist ads.

1. YARN, YARN, YARN! [A reference to "yarning" in *Ecofeminism Now!*: the weaving of intricate yarn webs through trees and around earth-destroying equipment to protest logging.]

Representations

Issues of representation are urgent for oppressed groups because representations act as propaganda for dominance; transforming and resisting oppressive representations is part of the struggle toward liberation. From applying graffiti to sexist and racist ads to the anonymous work of the Guerilla Girls challenging the racism and sexism of the contemporary art world, interactions with representations have been an integral part of feminist activism.

1960. Jane Goodall and her mother Vanne arrive at Gombe Stream Chimpanzee Reserve in western Tanzania.

John Berger's work in *Ways of Seeing* (1972) and *About Looking* (1980) raised issues about looking at women and nonhuman animals. "Why Look at Animals?" became a foundational essay in considering the way humans represented nonhumans. Berger proposes that the animal may have been the first metaphor and concludes his essay considering the role of animals in zoos: "The zoo cannot but disappoint. The public purpose of zoos is to offer visitors the opportunity of looking at animals. Yet nowhere in a zoo can a stranger encounter the look of an animal. At the most, the animal's gaze flickers and passes on. They look sideways. They look blindly beyond. They scan mechanically." Meanwhile, in *Ways of Seeing*, Berger identified how *men look at women and women watch themselves being looked at.* For many feminists in the 1970s and 1980s, the decades when Berger offered these insights about representation, issues of representation and sexual violence meant interrogating pornography. Susanne Kappeler's *The Pornography of Representation* (1986) brought the issues together, especially in a chapter indebted to Berger and extending him: "Why Look at Women?" Kappeler and others argue that, like zoos, pornography offers only a subject-object relationship, rather than one of intersubjectivity.

Resisting stereotypes and examining harmful depictions unites many social change movements. Michael Harris in *Colored Pictures: Race and Visual Representation* (2003) points out that "societies articulate their values and social hierarchies *visually* in many grand and subtle ways" (11). He argues that "blackness, unlike Jewishness or Irishness, is primarily visual" (3). He considers black artists' responses to racist imagery, which have included recontextualization and reappropriation. Drawing on visual studies, ethics, and theories of cognition, Rosemarie Garland-Thomson's *Staring: How We Look* (2009) provides ways to think about visuality and the body, helping to forward the discussion of representations of disability. Garland-Thomson points out that "In late capitalism, the predominant form of looking, the mass exercise of ocularcentricity, is what we might call consumer vision" (29). She refers to Lizabeth Cohen's suggestion that "one central task of citizenship in our era is consuming" and in that cultural context virtually everything and everyone becomes a commodity. Representations of women and men of color, shown with non-human animals or as non-human animals, contributes to commodification—both the selling of commodities, often the dead bodies of animals used for food, and the commodification of the disenfranchised human's body.

In the 1990s, Brian Luke examined the role of imagery in perpetuating oppressive attitudes in his consideration of the yearly *Sports Illustrated* swimsuit issue. Luke points out that "The *Sports Illustrated (SI)* swimsuit issues are recent examples of this cross-imagery between erotic sport hunting and predatory heterosexuality." In addition, "Race is heavily exploited in the *SI* pictorials. There are five pictures of women of color in the 1996 issue, and in each case the model is dressed in an animal print bikini and/or a suit with a native African motif." Luke concludes, "women, animals, and people of color all share a common status as objects placed on display for the white male viewer's erotic entertainment" (1996).

Early on some ecofeminists chose the medium of slide shows to problem-atize representations of women and the other animals. In the 1980s, Marti Kheel began creating a slide show that at first was called "The Re-Presentation of Women and Animals" and then "Women, Nature, and Animals through an Ecofeminist Lens." Eventually, in the 1990s, as others showed it, it became known as the "Feminists for Animal Rights Slide Show." Drawing upon an (unfortunate) wealth of cultural images, the slide show illustrated how the dominant culture presents, in her words, "women and animals as wild, demonic beings that must be subdued, and as inanimate objects that exist to serve 'man's' needs." Pulling images from pornography, mythology, popular culture, art, children's storybooks, as well as real-life pictures of animal abuse, the slide show was "a visual dissection of the patri-archal mind," tracing the images of women and animals from ancient history through the era of Cartesian science.

Though *The Sexual Politics of Meat* contained few images, in response to the book, readers began to send Adams other examples of what she calls the animalizing of women and people of color, and the feminizing, sexualizing, and racializing of domesticated animals. From this, she created *The Sexual Politics of Meat Slide Show.* She notes the fetishizing nature to certain images; the use of irony in the depiction of the heterosexual male as a meat eater, as if through the irony an indemnification of the message is provided. She also encourages those who see the slide show to ask whether the re-presentation of negative, oppressive, or violent images participates in the very visual culture that it attempts to deconstruct. Or can negative images that re-inscribe oppressive attitudes be contained, by being shrunk, or surrounded with text, or otherwise dislocated from their original context?

The circulation of oppressive images without an interpretive or explan-atory frame commits discursive violence that allows for a material form of violence. Experiencing the slide shows, viewers are urged to look with resistance, to recognize that images can't be unanchored from their referent—to refuse to consume the images on their own terms.

Over the past 20 years, developments in animal activism have raised new issues in representation. In response to undercover campaigns that

1968. Feminists and civil rights activists protest the "Miss America" contest. Feminists protest it as a cattle show.

expose what happens in factory farms, states are passing "ag-gag laws" that make it illegal to film on a factory farm or to misrepresent oneself to obtain access to a factory farm. Such laws recognize the power of the image in dispelling the consumer's nonchalance in consuming animal products, alerting them to the fact that this "commodity" had a life before ending up on a plate or before her milk became someone's coffee creamer. The ag-gag laws have been passed to prevent a specific kind of seeing, a specific kind of representation, a representation of the real lives and deaths of farmed animals.

Within the animal activist community, there has been discussion about whether people should be encouraged to watch videos that show the gory truth of animal agriculture. Labeled as 'bleeding Jesus' pictures by some, encountering such representations can be very upsetting. Advocates say, "Good, people need to be aware of the truth of what is going on." They know that these representations have been effective in motivating people. Some say, "We don't want to hide these images, but we should not simply use them to shock and upset." They know that these representations have also paralyzed some people, and desensitized them. The debate about what and whether representations catalyze positive change in humans toward their victims erupts each time PETA stages what it sees as a necessary act in advocating for animals and its critics view as egregious exploitation.

Feminists have used a lot of ink critiquing PETA's various "Naked" campaigns. It often isn't clear precisely what message they are selling— women look sexier if they don't eat meat or wear fur and men are better sexually or will get better sex from women if they don't eat meat? Sex, traditionally, sells, but, as some feminists have argued, the message of care, compassion, and justice gets lost in the exchange. In a response to complaints about the offensiveness of some of their campaigns, PETA's website states: "PETA does make a point of having something for all tastes, from the most conservative to the most radical and from the most tasteless to the most refined."

Herein lies one of the problems with single-issue campaigns—the larger political context in which the messages are presented is ignored. Everyone's "tastes," particularly those of the sexist, racist, or homophobe, should not be catered to simply because those people otherwise support the liberation of animals. Attitudes of human exceptionalism, entitlement, and disrespect play a central role in the social rejection of the idea that other animals matter, just as attitudes of white male superiority and entitlement play a central role in the perpetuation of racism and sexism.

Artists such as Kathryn Eddy, Sue Coe, New Zealand-based Angela Singer, Chicago-based Nancy Hild, and many others are demonstrating how

Figure 1.4 Kathryn Eddy, *Pig Blindness,* mixed medium collage on wood, 20 × 80, 2013. The artist writes: "What happens when animal ethics and popular culture collide? We live in a society that is obsessed with looking at celebrities and fashion, and fashion magazines provide the perfect venue. At the same time, we disavow the serious abuse of farmed animals caught up in the web of our food industries. As an artist whose work explores the human–animal relationship, this series points out our obsessive fashion voyeurism and our blindness towards our farmed food animals. A slaughterhouse employee remarked that his job wouldn't be so bad if he didn't have to look at the animals because they have such human eyes. By giving the pigs human eyes and elevating them to celebrity model, this work reverses the gaze from the animal to the human, thus redirecting and opening our eyes, hearts and minds to discussion and change." © Kathryn Eddy. Used by permission.

to artistically represent issues of human–animal relations without requiring any more deaths. In a series of books, Steve Baker has considered representations of animals: *Picturing the Beast: Animals, Identity and Representation* (1993), *The Postmodern Animal* (2000), and *Artist Animal (Posthumanities)* (2013). In his latest book, Baker examines the work of contemporary artists who directly confront questions of animal life, treating animals not for their aesthetic qualities or as symbols of the human condition but rather as beings who actively share the world with humanity. The creation of the National Museum of Animals and Society in Los Angeles, California and their promotion of innovative exhibits, will also allow for greater analysis of non-violent, intersectional representations.

1969. Stonewall Riots occur in New York City. In response to raids by the police, the gay community protests, paving the way for the Gay Liberation Movement.

Dominant discourse and disappearances

Reflections on the development of the animal protection movement usually tell the story of its beginning with the publication of Peter Singer's *Animal Liberation* in 1975. The early feminist concern for other animals is generally overlooked. While Singer's work has certainly had tremendous impact, as has the work of other theorists like Tom Regan and Gary Francione, identifying animal liberation as having "fathers" such as Singer and Regan and no "mothers" has generated a few problems. Both Singer and Regan, in an effort to legitimize concern for other animals within the tradition of analytic philosophy, relied too heavily on a division between reason and emotion and the gendered associations that attend to each. This framing foregrounds their approaches to thinking about our ethical obligations to other animals and relegates to the background feminist relational commitments to empathy and care. Sadly, ecofeminism was often seen not only as a side concern, but more problematically, ecofeminists were mistakenly thought to maintain the masculine reason/feminine emotion binary that mainstream theorists were promoting. Exposing dualistic frameworks operating in oppressive situations did not mean that ecofeminists valorized the non-dominant parts of the dualism nor viewed the characteristics of the non-dominant part as "natural." In arguing relationally and developing a care tradition in animal ethics, ecofeminists were challenging, not accepting, the essentializing structure of the division between men as rational and women as emotional.

At the turn into the twenty-first century, the persistent mischaracterizations of ecofeminism as essentialist continued to surprise many of those working in the field. This mischaracterization provided another way to exclude ecofeminist work that challenged the dominant discourse. Enabled in part by this continuing misreading, another problem became obvious, the failure to credit ecofeminism's influence. The emerging field of Animal Studies grows out of, at least in part, feminist and ecofeminist theory. But this background and history is often ignored or distorted in many discussions of animal studies. The result is not only the disappearance of ecofeminism, but the appropriation of ecofeminist ideas in which embodied authorship disappears.

Gaard's "Ecofeminism Revisited: Rejecting Essentialism and Re-Placing Species in a Material Feminist Environmentalism," initially presented at Wesleyan's 2011 Sex/Gender/Species conference, provides an intervention against the erasure of ecofeminist theory. Gaard writes: "These omissions in ecocritical scholarship are not merely a bibliographic matter of failing to cite feminist scholarship, but signify a more profound conceptual failure to grapple with the issues being raised by that scholarship as feminist, a failure made more egregious when the same ideas are later celebrated when presented via non-feminist sources" (3). She raises important concerns about name-calling (ecofeminists are referred to as "strident," "anachronistic," or "parochial") and suggests that this tactic is not only destructive of scholarly community, "anti-feminist name-calling may indicate the speaker's own lack of familiarity or even hostility to feminist perspectives" (17).

Sadly, it isn't just the dismissal of ecofeminism based on misunderstandings and hostility to the view that is troubling, but some feminist theorists have strategically decided that since ecofeminism continues to be mischaracterized it may be best not to align one's own work with it.

Gaard suggests that "the intersectional analysis of nature, gender, race, class, species, and sexuality is not confined to an essentialist definition of feminism or ecofeminism, but rather offers a strategic conceptual approach toward bringing about the social justice, economic and ecological democracy needed to solve environmental crises in the present moment" and hopes that her analysis may shed new light on the value of ecofeminist praxis.

Susan Fraiman was also at the Sex/Gender/Species conference at Wesleyan and developed her work questioning "the framing of animal studies in opposition to emotionally and politically engaged work on gender, race, and sexuality." She provides an important critique of the gendered omissions and distortions in the story of the beginnings of critical animal studies in her essay "Pussy Panic versus Liking Animals: Tracking Gender in Animal Studies." She writes:

> If Derridean animal studies seems poised to corner the contemporary market, I am troubled in part by its revisionary history—the way an origin story beginning in 2002 [Derrida's "The Animal that therefore I am (more to follow)"] serves to eclipse the body of animal scholarship loosely referenced above, dozens of books going back some forty years, long before Derrida's essay was brought to the attention of English speakers. Much of this pioneering work was by women and feminists—a significant portion under the rubric of ecofeminism—and all of it arose in dialogue with late-century liberation movements, including the second wave women's movement. (92)

Both Fraiman and Gaard note how feminist and ecofeminist insights and theory have been both discredited *and* appropriated. Interestingly some of the most prominent work being done in animal studies is critical of much the same structure of Singer and Regan's work as the ecofeminists before them, though it is leap-frogged over. Writing of Cary Wolfe's *Animal Rites,* Fraiman suggests it "effectively authorizes its critique of speciesism

1971. Frances Moore Lappé's *Diet for a Small Planet* published, prompting many to become vegetarians.

and models for contemporary animal studies by means of a revamped genealogy—one skewed to privilege Derrida and disregard the groundwork laid by ecofeminism" (103).

Moving the work forward

Despite various obstacles, ecofeminists have continued their intersectional analyses in the US and around the globe.

Thoughts of Ecofeminism in Israel
Shira Hertzanu

As an Israeli feminist and animal rights activist, I believe that the oppression of women and animals should be understood as part of the collective nationalist atmosphere. The lack of compassion towards Palestinians intersects with the lack of compassion towards animals. The non-recognition of women's rights and the feminist movement relates to the misunderstanding of the Palestinian aspiration for liberation. Indeed, a wide eco-feminist struggle to smash patriarchy in all its forms is due.

One of the methods used by Israeli soldiers and civilians to demean Palestinians is comparing them to animals, as patriarchs do to women. That is, in order to justify their claimed superiority, Israeli nationalists aim to lower Palestinians to an animal level. This is carried out either by verbal comparison or by actual treatment. Palestinians are compared to living animals, who are associated with weakness, lack of sophistication and barbarism.

Once an animal dies, he or she becomes something of value. After taking the form of meat, animals are no longer associated with weakness, but rather with strength. Throughout history, meat, as well as other animal-derived products, has been a central part of the combatants' diet. While Israeli soldiers receive unlimited amounts of food—except for special trainings in which there are limitations, the amount of food in military bases is far above the soldiers' needs, and excess food is thrown out daily—Israel has set firm restrictions over the food commodities allowed inside the Gaza Strip, which is under a de-facto siege. Therefore, by limiting the food intake, Israel aims to physically and morally weaken the Palestinians living in Gaza. Israel in fact uses meat, the strong (dead) animal, to weaken Palestinians, who are compared to weak (living) animals.

The Israeli society degrades not only Palestinians, but also human rights activists who support them. Activists of all genders and sexual orientations are accused of betrayal and of jeopardizing national security. Women and members of the LGBT community are also insulted based on their gender and sexual orientation. While there may be some inner logic in labeling activists as "traitors" (as they are anti-nationalists), referring to women activists as "whores" has no actual basis, and is done for the sole purpose of humiliation. Also, women (and LGBTs) who are arrested for political acts are often sexually harassed by the soldiers and policemen who arrest them.

Shira Hertzanu is an Israeli animal and human rights activist, active in different radical and social struggles. Shira is a member of Kasamba: Rhythms of Resistance, Tel Aviv—a group that uses percussion as a political tool, and is a board member of Coalition of Women for Peace, a feminist NGO to end the Israeli occupation of Palestine. She is a lawyer, and focuses on human and animal rights in her legal work.

Figure 1.5 Palestinian women at a checkpoint. An example for the animal-like treatment can be seen in some checkpoints, which are built using fences and cages, crowding Palestinians that wish to pass through, producing images that resemble those of animals in factory farms. Photograph © by Tamar Fleishman. Used by permission.

Globally, ecofeminists have continued to build on and extend their intersectional activism. lauren Ornelas, founder of Food Empowerment Project, writes:

> Many of us who work on animal issues have been able to do so because we are privileged enough to not have to deal with environmental racism where we live—we're not fighting against environmental toxins in our communities, we don't have family members dying of cancer due to these toxins, and most of us are not simply struggling for survival. From vegan chocolate coming from West Africa that is laden with slavery and the worst forms of child labor to the plight of farm workers in the fields who pick our produce, we all need to recognize the impact that our food choices have on others. ... There are approximately 3 million people who work in agriculture in the US. A majority of these are people of color who live in rural communities. (2012)

Their work of bending, crouching, and lifting is taxing to their bodies, while at the same time they are exposed to agricultural chemicals. Meanwhile, their pay is often below the minimum wage, they receive no health care or benefits, and their children often work along with them in the field, because as Ornelas points out, "it is legal for children, who are not yet even teenagers, to spend their days picking produce."

Ornelas has pioneered a creative dialogue about environmental racism. Food Empowerment Project has explored food deserts, in which people lack access to fresh fruits and vegetables and there is an overabundance of liquor stores, convenience stores, and fast food restaurants. Food Empowerment Project has also explored food availability in Santa Clara County, California; they concluded that "It was almost impossible to find meat and dairy alternatives in communities of color, yet a majority of people of color are lactose

1971. Birute Galdikas begins her work with orangutans in Tanjung Puting Reserve in Indonesian Borneo.

intolerant." Next, Food Empowerment Project conducted focus groups, all in Spanish, in the most impacted communities and found that the cost of and distance from healthier food items were barriers to using them in food preparation.

While working to include intersectional analysis within animal and environmental activism, it is hard not to notice important lacks within the animal activist community. Many ecofeminists have discovered that this community has been particularly resistant to addressing issues of oppression within its ranks, including racism, sexual harassment, and sexual violence perpetrated by one activist upon another and ignored by the white, male "leaders" that continue to be at the fore of the movement over 30 years after the early criticisms of exclusion and privilege were raised.

One explanation for this failure is that activists tend to be highly skeptical of the legal system in general and in particular as a venue to address domestic violence. They view the carceral system as racist and inhumane and believe that it neither provides redress for survivors nor helps perpetrators to recognize the harm they've caused. Instead, they prefer models where perpetrators voluntarily agree to participate in a process where they are confronted with their actions and encouraged to take steps outlined by abuse survivors to address the survivors' needs and also to address their own behaviors. Perpetrators may be asked to seek counseling to address their abusive behavior, seek treatment for substance abuse if their abuse is linked to use of alcohol or another drug, to cover the cost of medical treatment or counseling that the survivor had to seek as a result of the abuse, to stay away from events the survivor may wish to attend in order to allow the survivor to feel safe, or any of a number of other measures aimed not at harming or punishing the perpetrator but at providing healing for both parties. This approach recognizes that many abusers are themselves survivors of abuse and aims to stop the cycle of responding to harm by causing harm.

When perpetrators refuse to be accountable for their actions, activists who support this approach are still reluctant to rely on police and courts as a solution. Instead, operating on the premise that a person who refuses to be accountable for abuse is likely to abuse again, they ask the community not to welcome the abuser into its spaces, gatherings, and conferences. This allows the survivor some modicum of safety. But in the absence of community support, it is almost impossible for such approaches to be effective.

The animal movement and the environmental movement, like many social justice movements, are also reluctant to engage in critical analysis of themselves. Anastasia Yarbrough recently reported on a national conference in which racism, sexism, and homophobia and transphobia were far too common. She notes that the movement continues to use "the struggles of people of color and women as a means to motivate" without knowing much about the lived experiences of oppression and suggests this tactic is another form of exploitation.

> After listening to white affluent plenary speakers repeat the same comparisons of racism and speciesism over and over again without the slightest insight as to how racism operates in America, their comparisons made me feel more violated and misused than usual. How dare they talk about racism in this superficial way when they're not even willing to address how it's unfolding at this conference? How dare they use the plight of my people to justify a righteous enterprise for animal rights? How dare they use animals and people of color in this way?
>
> From beginning to end, this conference was not rendered a safe space for people of color, women who experienced sexual assault at the conference prior and received no justice, and lgbtq people. Harassment and violence seem to be the order of the day...[4]

Often efforts to seek restorative justice and demand accountability have resulted in reinforcing the power of perpetrators, who've effectively employed the same terror tactics, manipulation, and domination to their backlash strategy as they did to their initial ignorance and abuse.

Ecofeminist theory and practice, in working intersectionally, can provide a means for addressing these problems for those willing to reflect upon them. Theoretical work in ecofeminism identifies the interconnected structures of normative dualisms, highlights the ways that such dualisms facilitate oppression and misrecognition, and draws out both conceptual and practical connections between injustice towards non-dominant individuals and groups. In practice, ecofeminists work in solidarity with those struggling against gender oppression, racism, homophobia and transphobia, environmental injustice, colonialism, speciesism, and environmental destruction. In both theory and practice, ecofeminists imagine different social relations are possible and encourage work to achieve peace and justice for all.

Acknowledgments

We would like to thank Greta Gaard for the idea to present these ideas in a visually accessible way. We especially thank Rosemary Ruether, Batya Bauman, Karen Davis, and Kim Stallwood for allowing us to highlight some of their presentation in boxes. We also thank lauren Ornelas and Adam Weissman for their important presentations at the 2012 Wesleyan conference, some of which clearly inform the ideas in this chapter. We would like to acknowledge Shira Hertzanu for responding to our request to write about the politics of animals in Israel. Finally, we thank the many, many activists and scholars whom we do not directly cite here who have been engaged in ecofeminist praxis.

Notes

1   This was a topic that received public attention during Anita Hill's testimony in the Clarence Thomas confirmation hearings. Crenshaw herself noted: "rape and other sexual abuses were justified by myths that black women were sexually voracious, that they were sexually indiscriminate, and that they readily copulated with animals, most frequently imagined to be apes and monkeys" (Crenshaw 1992, 411).

2   See the interview posted at the VINE blog here: http://blog.bravebirds. org/archives/1568

3   http://www.theveganrd.com/2010/09/review-of-the-vegetarian-myth. html

4   Yarbrough's full reflection on the conference can be found at her blog here: http://animalvisions.wordpress.com/2013/07/09/ recalling-the-animal-rights-conference–2013/

part one

Affect

Compassion and Being Human
Deane Curtin

2

Introduction: Compassion vs. rights

Ecofeminists share an assumption that the care/empathy/
compassion approach to ethics, one that highlights the
centrality of affect, is somehow more fundamental than an
abstract ethic of rights. An ethic of care is certainly more
inclusive since it values the diverse ways that women and men
tend to organize their moral experience. Also, since it does not
begin with the assumption that ethics should be built on a feature
that is uniquely human—*human* reason—it is less anthropocentric. But is it more *fundamental* in the sense that care goes
deeper in our moral experience than rights? To answer this
question we need to investigate the relationship between affect
and rationality in a relational ethic of care.

First, so we know what we're talking about, some clarification of terms.

By the rights approach I mean the view of moral functioning
that is central to political liberalism, the philosophical view
shared by otherwise disparate philosophers such as Hobbes,
Locke, Mill, and John Rawls. On this view, individual rights are
basic. Liberals are, therefore, committed to a strongly individuated rational self, the morally autonomous self as the holder of
these rights. In turn, rights always imply duties. I cannot have a
right to freedom of speech if I don't also respect your freedom.

1973. Connie Salamone's "Feminist as Rapist in the Modern Male Hunter Culture" published in *Majority Report*, the NYC feminist newspaper.

Liberals usually distinguish between positive and negative rights. Most of the central rights in the Western tradition are negative. As Justice Brandeis famously said, the most important right is "the right to be let alone."[1] That is, the right to privacy is basic in liberalism. Freedom of religion, for example, is basically the right to privacy, to be "let alone," in making decisions about religious preferences.

This is why the distinction between the public and private spheres of life is so central to liberalism. The public sphere should be minimal, just enough to establish basic fairness, behind the veil of ignorance in Rawls's famous variant, so the private sphere can be maximized. I am free if society leaves me alone to choose substantive goods for myself.

Positive rights, on the other hand, tend to be more localized. They operate within the context of functioning institutions. The right to an education, for example, makes a positive claim on society. It functions only within the sphere where there are educational institutions, and where there has been an agreement between state and citizenry to provide education. Positive rights are special cases, so most of what I will say about compassion vs. rights pertains to negative rights.

I generally prefer the word "compassion" to care or empathy. For me, compassion is a *developed* moral capability, whereas care or empathy are closer to the natural capacities that make compassion possible. Humans, and many other animals, naturally have empathy for the suffering of others. Compassion, on the other hand, is a cultivated aspiration to benefit other beings.

I use the word "compassion" in the way that the Dalai Lama articulates it in his recent book, *Beyond Religion*:

> ... although compassion arises from empathy, the two are not the same. Empathy is characterized by a kind of emotional resonance—feeling with the other person. Compassion, in contrast, is not just sharing experience with others, but also wishing to see them relieved of their suffering. Being compassionate does not mean remaining entirely at the level of feeling, which could be quite draining. After all, compassionate doctors would not be very effective if they were always preoccupied with sharing their patients' pain. Compassion means wanting to do something to relieve the hardships of others, and this desire to help, far from dragging us further into suffering ourselves, actually gives us energy and a sense of purpose and direction. When we act upon this motivation, both we and those around us benefit still more. (2011, 55)

This passage highlights two issues that I will explore here. One is that compassion is not only an emotional response; it blends reason and feeling together. The other issue is that, in blending reason and feeling, compassion becomes more resilient than empathy. This resilience is what makes an ethical practice possible, and it is the core of the defense of compassion against the charges of political liberalism.

Even this brief sketch of the two approaches to ethics points toward major differences between them. Whereas liberalism only makes sense on the assumption of a strongly individuated, autonomous self as the rights-holder, compassion grows out of a relational self. Whereas rights and duties are reciprocal between rights-holders, and therefore the moral domain is confined to those who *can* reciprocate (Kant's rational "persons"), true compassion is defined by not being reciprocal. In fact, if we discover that a person's motive was based on the expectation of repayment we take this as evidence that the motivation was not compassionate.

Furthermore, since the divide between self and other is so central to liberalism, a core issue is egoism vs. altruism. The liberal picture is: should I rest content to benefit *this* autonomous agent, or am I obligated to leap over the self/other divide and help others? Given the relational self, the egoism/altruism divide simply does not make much sense in the compassion perspective. In general, as the Dalai Lama says at the end of the quotation, helping others is also helping oneself. Ethics can then be seen as a practical collaboration with others, even others with whom we disagree. It is not a zero sum competition.

Finally, I'll suggest that whereas liberalism sees the task of ethics as providing rules through which we can distinguish good from evil, compassion sees the basic task of ethics differently. One model for an ethic of compassion is health.

A final clarification: I will not argue that rights talk should be eliminated from moral discourse, only that compassion is more basic. Sometimes we hear that liberalism is so dear to Western moral intuitions that applying it universally amounts to a kind of liberal imperialism. Although I am somewhat sympathetic to this view, in the end I maintain that rights are useful in certain contexts.[2] So understood, they are complementary to an ethic of compassion.

In what follows, then, I discuss the contrast between negative rights and compassion, arguing that compassion is basic. It is the fundamental way in which we relate to other beings in the moral sphere of our lives. Rights have their place, but they are secondary. Although I cannot argue this fully here, I believe justice serves compassion. That is, the justice of any legal system we can endorse consists at least partly in that the justice it renders is compassionate.

1975. Peter Singer's *Animal Liberation* published.

Arguments against compassion

As I read Western philosophy, there have been two basic arguments against the primacy of compassion, one epistemological and the other ethical. The first is familiar. It is endorsed most prominently by Plato. Rationality is universal, objective, impartial, and since we want ethics to guide behavior it cannot be subjective. It must depend on objective reason rather than subjective emotions. Compassion depends on the emotions.

The ethical argument is in some ways more interesting and challenging. It comes from the Greek and Roman Stoics, from Spinoza, and, in its most extreme form, from Nietzsche. This view holds that compassion actually leads to destructive emotions and states of character, and therefore to immoral action. The Stoic version goes like this: any attachment to apparent goods outside oneself leads to suffering, since, by definition, we cannot control external events. Therefore, the key to a happy life is to depend only on oneself to the extent that one is rational. Compassion is an emotion that attaches us to outside goods. Therefore, it is contrary to a moral life. The Stoics add that in rejecting external goods we also cut off the main cause of human conflict, namely conflict over external goods: fame, fortune, etc. So rejecting compassion can lead to a peaceful life.

Spinoza, Kant and Nietzsche pick up on the Stoic idea that compassion is a weakness that diminishes the possibility of a moral life. It does so by demeaning those who receive our compassion. Kant referred to compassion as an "insulting kind of benevolence" (Nussbaum 2001, 378). Compassion is a form of pity in which those who are superior look down upon recipients. It rewards moral weakness.

Both of these arguments depend on the assumption that there is a categorical divide between reason and emotion. Reason is objective, dispassionate, and universal. Emotion is subjective, flighty, and dangerous to self and others. While there are substantive philosophical differences here, we also need to be alert to the possibility that Western philosophers assume culturally specific definitions of compassion. Many dictionaries *define* compassion in terms of pity. This is a mistake, a cultural bias, that can be corrected by looking at compassion across philosophical traditions.

What makes morality possible?

There are deep biological reasons for the priority of compassion, reasons having to do with who we are in the world. Evolution has made empathy, and therefore compassion, possible for human beings. In the 1990s researchers at the University of Parma located the so-called mirror neurons in the F5 part of the brain. These neurons make empathy possible. So, if I yawn, many others in the room are likely to start yawning, even if they aren't tired. Or, when a mother trying to feed her baby opens her mouth in sympathetic response to the baby's mouth opening, this empathic response is due to the functioning of mirror neurons. As philosopher of science, Thomas Metzinger, puts it, "Thanks to our mirror neurons, we can consciously experience another human being's movements as meaningful" (Metzinger 2009, 172).

Of course, this basic biology can be thwarted by upbringing. If children learn that they cannot trust the adults around them, if the world seems dangerous rather than welcoming, these parts of the brain do not develop. In fact, one line of research is exploring the question of how much human social pathologies can be traced back to developmental issues with mirror neurons. Not surprisingly, recent research indicates that mirror neurons may be more active in women than in men. So, the converse may also be true, that nurturing can make empathy more active.[3]

To the claim of being fundamental, therefore, it is arguable that we would not be human, we would not be social (and therefore moral) beings at all, without the basic genetic capacity to share meaningful experience empathically with others. It even makes possible the understanding that there *are* others, who, like ourselves, need to be treated with understanding and care. Empathy is perhaps a/the defining characteristic of being human. It is what has allowed us to be spectacular successes from an evolutionary viewpoint.

And yet, unlike the classical defining characteristic of being human that was supposed to wall humanity off from the rest of nature—Aristotle's "rational animal, or Descartes' "thinking thing"—taking empathy as basic exposes our interconnections to other beings. Contemporary science corrects the classic philosophical stance that human beings are categorically different from the rest of nature. Being human is a matter of degree rather than kind.

Mirror neurons were discovered initially in macaques,[4] and they are now known to exist in other non-human primates, in birds, and possibly in octopi. As Metzinger has pointed out, "The mirror-neuron story gives us an idea of how groups of animals—fish schools, flocks of birds—can coordinate their behavior with great speed and accuracy…" (173). Most interestingly from a moral viewpoint, the nonhuman others who interest us most, those who seem to "look back at us," are beings with developed mirror neurons: bonobos, elephants, dolphins. In short, mirror neurons are the biological foundation for a social self, a self engaged in a world that is meaningful. This trait is hardly unique to human beings.

1975. Rosemary Radford Ruether's New Woman, New Earth: Sexist Ideologies & Human Liberation published.

At a basic biological level, mirror neurons help maintain *homeostasis*, the basic biological ability to stay sufficiently balanced to survive. Neuroscientist Antonio Damasio puts it this way:

> What does it take for a living cell to stay alive? Quite simply, it takes good housekeeping and good external relations, which is to say good management of the myriad problems posed by living. Life, in a single cell as well as in large creatures with trillions of them, requires the transformation of suitable nutrients into energy, and that, in turn, calls for the ability to solve several problems: finding the energy products, placing them inside the body, converting them into the universal currency of energy known as ATP, disposing of the waste, and using the energy for whatever the body needs to continue this same routine of finding the right stuff, incorporating it, and so forth. (2010, 41)

In short, we share homeostasis with even single-celled organisms, but it shows itself in humans and many other biologically complex creatures through empathy. Sociability helps us survive.

For Damasio, the biological need for homeostasis is what generates "biological value" (48). Food and oxygen are nonmoral biological values required for survival. But Damasio further proposes that, "In an extraordinary leap, homeostasis acquires an extension into the sociocultural space. Justice systems, economic and political organizations, the arts, medicine, and technology are examples of the new devices of regulation" (26).

This is indeed an "extraordinary leap," from oxygen to justice systems. It raises the question of how biological empathy gets transformed into compassion.

Turning empathy into compassion

While we have the basic capacity to empathize, compassion as a *moral* commitment must be developed. Just because we are creatures with mirror neurons doesn't mean we develop the mature capability to act compassionately. In Martha Nussbaum's terms, the *capacity* must be developed into an actual human *functioning*, a *capability* (2000). This process involves cultivation of the moral feelings in which our basic empathy comes to be employed skillfully.

It is here, with the idea of moral skillfulness, that we meet the need to respond to the epistemological objection to compassion. If, as the critics assume, reason and emotion are categorically different, then empathy cannot be developed into compassion. Empathy is pure emotion, and therefore cannot guide moral development. This view, I believe, is a tragic mistake. In fact, when compassion flourishes, feelings and reasons blend and result in a moral practice.

I want to begin, again, with Damasio because I think he makes a useful distinction between emotions and feelings. "Emotion and feeling," Damasio says, "albeit part of a tightly bound cycle, are distinguishable processes" (109). Two important points here: emotions and feelings are tightly bound together in a cycle, and they are processes, not things.

He continues:

Emotions are complex, largely automated programs of actions concocted by evolution. The actions are complemented by a cognitive program that includes certain ideas and modes of cognition, but the world of emotions is largely one of actions carried out in our bodies, from facial expressions and postures to changes in viscera and internal milieu. Feelings of emotion, on the other hand, are composite perceptions of what happens in our body and mind when we are emoting. As far as the body is concerned, feelings are images of actions rather than actions themselves; the world of feelings is one of perceptions executed in brain maps ... (109)

It would be easy to misrepresent this subtle distinction, but I think the main points are clear. An emotion can result from a temporary disruption of homeostasis, for example, fear at the prospect of danger. The "fight or flight" mechanism that evolution has embedded in us is a basic way of trying to return to balance. There is certainly a cognitive element here. At some very basic level, there is a judgment: "Danger is present. Do something!" However, the judgment here isn't an extended process of reflection. Extended reflection can result in death. The cognitive element also cannot be disentangled from action. It is a process that immediately motivates action.

On the other hand, feelings are "feelings about emotion." They are more reflective, and the cognitive element is more obvious. On the level of feeling I may ask, and have the time to ask, whether I really should have fought in response to the danger. Was there an alternative to either fight or flight? Was there an opportunity for negotiation?

The same can be said about love and hate. Anyone who has experienced romantic love (a helpful evolutionary trait) knows how close this kind of love is to hate. One unfortunate word, or action, can radically reorient one's world from romantic love to hate, in an instant. But isn't there also a more reflective kind of love that is not so volatile in its response to unfavorable stimuli? A more reflective, compassionate, kind of love understands the causes and conditions of the situation, and isn't so quick to resort to "fight or flight."

1975. Carol J. Adams's "The Oedible Complex: Feminism and Vegetarianism" published in The Lesbian Reader.

Feeling, then, is emotion modified and cultivated. Emotions are immediate, bodily, social, empathic reactions. Feelings evolve out of emotions as a moral practice develops. We might say that emotions are nonrational (not, as they are often depicted, irrational). Feelings, on the other hand, are the paradigm in human experience of the integration of reason and feeling, thinking and doing.

The problem with the epistemological objection is that it doesn't recognize the distinction between emotions and feelings. Emotions are nonrational reactions; they are not irrational, as the objection would have it. Feelings about emotions have strong cognitive elements. But, unlike the abstract liberal view of moral rationality, feelings arise from and are directly engaged with action.

Compassion is a cultivated feeling about emotion. It is a place where how we feel, how we think, and how we act come together. In other words, compassion is a cultivated *practice*, not an isolated, rational judgment about the world. It is a deep, ongoing pattern of engagement. As a practice rather than an isolated rational judgment, it can come to look spontaneous. This connects with ideas of virtuosity. Just as a great violinist practices for countless hours so that in concert the performance appears completely spontaneous, so it is with compassion. It becomes who we are, through countless hours of practice.

Defining compassion

Part of the problem in talking clearly about compassion is that it is undoubtedly affected by social context, and by issues of race, gender, and class. Western philosophy, as I have noted, simply defines compassion differently from, for example, the Buddhist tradition as seen in the quotation from the Dalai Lama. So, it might help to clarify what I mean by considering the work of two philosophers whose work on compassion comes very close to mine, Martha Nussbaum and Lori Gruen.

Martha Nussbaum has written a monumental account of compassion, thinking primarily in the Western tradition coming out of Aristotle. There is much in her account that seems right. She describes emotions (what I call feelings) as "intelligent responses to the perception of value," and she continues:

> If emotions are suffused with intelligence and discernment, and if they
> contain in themselves an awareness of value or importance, they cannot,
> for example, easily be sidelined in accounts of ethical judgment, as so often

they have been in the history of philosophy. Instead of viewing morality as a system of principles to be grasped by the detached intellect, and emotions as motivations that either support or subvert our choice to act according to principle, we will have to consider emotions as part and parcel of the system of ethical reasoning. (2001, 1)

I agree.

However, as she goes on to define compassion, I have concerns: "To put it simply, compassion is a painful emotion occasioned by the awareness of another person's undeserved misfortune. Compassion, in some form, is also central to several Asian cultural traditions" (2001, 301). The "in some form" here covers a lot of ambiguity. In at least some Asian traditions I don't think compassion is a painful emotion, and the application of "undeserved misfortune" is not clear. Think back to the Dalai Lama's mention of "compassionate doctors" who "would not be very effective if they were always preoccupied with sharing their patients' pain." His point seems to be that a doctor best serves her patient by *not* feeling the painful emotion that is associated with empathy. Nussbaum's subtle account of compassion is designed to illuminate an Aristotelian, essentially tragic, view of life. Her compassion is the feeling we might experience from a great portrayal of Oedipus as he blinds himself, having discovered that the person with whom he has had sex is his mother.

In the South Asian Buddhist tradition, one of the three main causes of *dhukka* (suffering) is the very fact that we are alive. So we cling to life despite the fact that our basic condition is to be mortal. Could this condition be accurately described as "undeserved misfortune"? Our mortality is just a fact, to which we need to adapt. It is not undeserved, as if there is some other option. So it shouldn't be thought of as a misfortune. In some sense, it is our great "fortune" to have lived, although living is impossible without suffering.

Typical of Greek tragedy, Nussbaum's view finds compassion in places where things could have been different. That's the "tragic flaw" that makes tragedy possible. Things could have been different for Oedipus. In contrast, I think there's an element of Stoicism in some strains of Buddhist compassion: compassion requires us to develop skillfulness in adapting to reality. Only when we adapt to reality is there promise of relief from suffering.

Continuing with Buddhist terms for a moment, this tradition holds that wisdom and compassion are the same thing, just viewed from two sides. Without wisdom, compassion is easily misguided. Compassion requires wisdom concerning the true causes of suffering. Conversely, without compassion, wisdom is coldly analytical. Compassion involves a fusing of judgment and feeling about such important life situations.

I want to cite just one historical example of the use of reason, or wisdom, in shaping compassion, because I think it is instructive. The great Buddhist philosopher of the eighth century, Śāntideva, wrote the source text for most thinking about compassion in the Tibetan Buddhist tradition. Just to give some idea of his influence, the Dalai Lama has said that, "Insofar as I understand anything about compassion I learned it from Śāntideva." Not a bad book blurb.

Śāntideva was deeply concerned with anger, because anger is overwhelming; it occurs in a flash, and it kills compassion. One of his most celebrated pieces of advice to soothe anger goes like this: Suppose you have an enormous problem that is causing you great distress. Śāntideva counsels, "If there is a solution, then what is the point of dejection?" On the other hand, "What is the point of dejection if there is no solution?" (Śāntideva, Crosby, and Skilton 2008, 50). In other words, *stop wasting time with anger!* If there's a solution to your problem, get about solving it. There's no reason for dejection. If your problem really can't be solved there's no reason for dejection either, since there's no alternative. It's just the way things are. On either alternative, anger is a waste of time.

Now, whatever you think of this argument, I want to note that it is a rational argument. It has premises and a conclusion. But I think we would be misled if we think it is just an abstract, rational argument. Because, certainly, Śāntideva wants us to change our feelings about emotion as a result of this argument. Its blunt alternatives are surely meant to make us feel that there is something silly about anger, despite its all-consuming heat when we are in its midst. He is using reason to defuse intense emotion, to dislodge it from our overheated attention. As part of a practice, that is, not as a single isolated rational argument, Śāntideva clearly believes this kind of appeal can be transformative. It can alter the way we feel.

If there's anything to this kind of argument, it shows that compassion is not mere liking. It is not even simply "putting yourself in someone else's shoes." That's empathy. Compassion is a practice in which one engages deliberately and reflectively, through which one becomes skillful in identifying the true causes of suffering in others and in ourselves, as in Śāntideva's analysis of anger. Compassion involves deep insight into the medicine that's needed for the disease. It results immediately in action, when possible, to relieve suffering, and encourage wellbeing. As Śāntideva's analysis demonstrates, however, compassionate insight also includes understanding of when it is *possible* to act. The second of Śāntideva's options is to change one's attitude, and thereby adapt to reality.

Despite these cross-cultural difficulties in talking about compassion, I certainly agree with Nussbaum that in compassion feeling and insight work together. In examining tragedy she also makes clear that compassion is invoked in cases that are central to life's meaning. A stubbed toe calls for empathy, but probably not compassion.

Philosopher Lori Gruen has also developed an account of ethics arising from the affects, which is particularly relevant and important here because her philosophical acumen is combined with experience involving nonhuman primates. Reflecting on many of the same concerns that motivate me—to bring the role of affects forward, to respect both similarity and difference in moral encounters, and to show that empathy can be moderated by reason— Gruen proposes a theory of "entangled empathy."

As she defines it,

> ... entangled empathy is a process whereby individuals who are empathizing with others first respond with a precognitive, empathetic reaction to the interests of another ... From these reactions, we move to reflectively imagine ourselves in the position of the other, and then make a judgment about how the conditions that she finds herself in may contribute to her perceptions or state of mind and impact her interests. These perceptions will involve assessing the salient features of the situation and require that the empathizer seek to determine what is pertinent to effectively empathize with the being in question. Entangled empathy requires room to correct empathetic responses. Entangled empathy involves both affect and cognition. The empathizer is also attentive to both similarities and differences between herself and her situation and that of the fellow creature with whom she is empathizing. She must move between her own and the other's point-of-view. (2013, 226)

It is this ability to move between one's own and the other's point of view that distinguishes the immediate, precognitive empathic response from fully entangled empathy. As she says, "Entangled empathy involves both affect and cognition." This distinction within empathy is motivated by the same concerns that cause me to distinguish empathy from compassion.

The distinction between empathy and entangled empathy allows Gruen to respond to a familiar criticism. Critics may allege that the accomplished torturer feels empathy with his victim. Empathy is required to decide what sorts of pain will be particularly effective. However, we don't think of torturers as paragons of virtue. So it appears that empathy isn't as intimately engaged with morality as Gruen and I suggest.

However, this criticism misses Gruen's point that entangled empathy goes back and forth between first- and third-person perspectives, negotiating both affect and cognition. As Gruen says,

1976. First (and only) meeting of the Society of Feminist-Vegetarians at Shandygaffs, a vegetarian restaurant in the Castro District of San Francisco, convened by Chellis Glendinning and Carol Adams.

> Empathy can be inaccurate and empathetic inaccuracies can take a variety
> of forms, both epistemic and ethical. Epistemic empathetic failures can
> involve overempathizing or incomplete empathizing. In cases in which
> over-empathizing occurs, the empathizer over-identifies with the emotions
> of another. This might occur between individuals who already have strong
> personal bonds—between friends, lovers, or between parents and their
> children or individuals and the other animals they care for. In cases of this
> type of empathetic inaccuracy, the empathizer exaggerates the emotional
> states of the one with whom she is empathizing. The empathizer may need
> to be less entangled. (2013, 227)

Again, this is an important point. Entangled empathy, since it balances affect with rational judgment, allows that there are cases in which the person who empathizes needs to be "less entangled" with the being with whom the empathizer is involved.

So, what hangs on Gruen's defense of entangled empathy and my defense of an ethic of compassion? It may be that Gruen and I are simply using different words for the same features of moral experience. I distinguish empathy and compassion partly because I depend on the distinctions found in neuroscience, whereas Gruen adopts terminology from experimental psychology. I also choose compassion because it is clearly understood within the Buddhist tradition, and I see no need to "reinvent the wheel." However, entangled empathy and compassion are certainly getting at most of the same points. In most cases we agree.

Nevertheless, I wonder whether the Dalai Lama, in the quotation I discussed earlier, is getting at something in addition to (not necessarily opposed to) the discussion of entangled empathy. What was he getting at when he spoke about compassionate doctors? Recall his remark: "After all, compassionate doctors would not be very effective if they were always preoccupied with sharing their patients' pain." If doctors constantly empathized with their patients' suffering they would quickly be ineffective. They would suffer from "empathy fatigue."

The Dalai Lama is saying something here that should provoke our sustained reflection. In cases of developed moral skillfulness, compassion sometimes needs to be *disconnected* from empathy. I wonder whether Gruen's description of entanglement fits these cases. Certainly it is a strength of her account that we can "negotiate" with empathic reactions: "Empathy can be inaccurate and empathetic inaccuracies can take a variety of forms, both epistemic and ethical." But we are still negotiating with empathy here. No doubt, much of our moral experience does involve such skillful negotiations, but what the Dalai Lama calls "universal compassion" seems different.

Imagine someone, like the Dalai Lama, who has devoted his life to compassionate engagement with those who suffer. He has seen more than 1.3 million of his own Tibetan people die at the hands of the Chinese government and military. Ninety-five percent of the monasteries in Tibet have been destroyed. It is illegal to speak Tibetan in school, and it is a crime to possess a shortwave radio, or a photograph of the Dalai Lama. Almost every day he meets personally with people who escaped from Tibet having been tortured. Most of those who escape know that they have paid the price of permanently losing contact with their families.[5]

Gruen's account obviously makes sense if we are thinking about the Tibetan side of this picture. One cannot hear the life stories of many Tibetans and not feel empathy. Still, a central feature of the Dalai Lama's teaching is that we should also act compassionately toward those who think of themselves as our enemies. He has even said repeatedly that we should value "enemies" more than friends because our enemies teach us compassion: "For a person who cherishes compassion and love, the practice of tolerance is essential, and for that, an enemy is indispensable. So we should feel grateful to our enemies, for it is they who can best help us develop a tranquil mind!"[6]

It is relatively easy to feel empathy toward family and friends. The real test of universal compassion is the ability to understand the causes and conditions behind even an "enemy's" abuse and have compassion for their suffering. What makes universal compassion so difficult, and important, is that it operates in the space where compassion is disconnected from empathy.

Or, consider another context where compassion must be disconnected from empathy—the case of women who survive abusers and somehow find a way back to health. First, of course, abused women must find a way to survive. They must get to safety and stop the immediate abuse. But, given the safety and space to reflect, doesn't a victim of abuse often find that misplaced empathy was at the core of what allowed a pathological relationship to continue? Overwhelming violence combines in a toxic mix with an occasional hint of what's represented as "love," and the relationship goes on. So, the second step toward health is to see what is really going on and stop the misplaced empathy. Health begins when there are no more negotiations.

Having stopped the negotiations over empathy, is it possible that something like compassion is required to establish real health? Without something like compassion, anger—toward self and other—will always remain at the center. Understanding the dynamics of abuse opens the possibility for the dissipation of anger. Health becomes available when empathy stops.[7]

It's also relevant here to draw out at least one dimension of this account of compassion for radical politics. People who have been marginalized by

1976. The first "gay rodeo" is held. By 1983, Feminists for Animal Rights engaged in protests of the event as upholding "men's most blatant attempt to celebrate the domination of nature."

systems of oppression might regard arguments for compassion with under-standable frustration: the oppressive social category through which culture defines the oppressed usually includes the expectation that the oppressed should be of service to others. We can imagine someone saying, here I struggle to claim an autonomous sense of self-worth, and Śāntideva tells me that morality requires that I put my self into service for *all* other beings. That doesn't sound like good news.

This concern deserves serious attention, but I think at least the beginnings of an answer are already evident, namely in the very distinction between emotions and feelings about emotions, and therefore the distinction between empathy and compassion. Emotions are subject to abuse. Feelings, however, by definition, require a degree of cultivated skillfulness. If we were to fully unpack what is involved in skillfulness in such cases it would certainly include the element of skill in overcoming oppression, which is a form of suffering.

What prevents being dragged down into empathy fatigue, therefore, is insight into the causes of suffering. This insight is what makes compassion universal. In the Buddhist analysis, there are three causes of suffering (the three poisons): attachment (greed), aversion (anger), and ignorance. I will not go into this analysis here. Discussions of this analysis are widely available. The point I want to make is that we need *some* understanding of the root causes of suffering if we are to act *skillfully* in response to suffering and avoid the paralysis of empathy fatigue.

So, the Dalai Lama's doctor example: a doctor would soon be fatigued if every case of the flu had to be treated as only an empathic response to each patient. There is something universal in the skillful treatment of disease, and that involves understanding that this is a case of the flu, as well as a unique being who is suffering with the flu.

Gruen talks very clearly about cases of "empathetic inaccuracy" in which the empathizer needs to be "less entangled." This is part of what I call skillfulness. I think we might agree that being less entangled in certain cases is a preventative to the caring fatigue. My concern is with what makes "empathic inaccuracy" inaccurate. It is not only that we have misunderstood the dimensions of a situation involving a concrete other. It may also be due to understanding the causes of suffering in general. Universal compassion depends on understanding such causes of suffering. Without this under-standing it is hard to know what counts as inaccurate in a concrete situation that calls for compassion.

The Stoic objection to compassion

We are now in a position to respond effectively to the second standard Western objection to compassion, that it is connected to pity, and therefore encourages moral weakness. Empathy, I have argued, is more basic to our moral experience than rights. It functions at the very source of what makes us human: sociability and meaning. Far from siding with emotions and neglecting reason, as the liberal account would have it, compassion is the model of how feeling and thinking are blended into a skillful moral practice.

In contrast, the rights that are most pervasive in our moral lives are merely *negative*. They are designed to apply to a partial sphere of our lives, the public sphere, and they apply only as constraints. They tell us what *not* to do. They do not pertain to, arguably, the most important, pervasive sphere of our lives, the supposedly private domain of the home. Liberalism says that the home is a "man's private castle." And yet, for most of us, the home is the place where we learn to be social beings. Long before introduction to the impersonal world of rights and duties, we learn to exercise empathy, appropriate trust, and connection.

The advocate of the moral objection might say at this point, this is exactly why compassion cannot be basic. Unlike the epistemological objection, which is based on a misunderstanding of the emotions, the moral objection raises serious questions about the limitations and dangers of an ethic based on compassion. Just because compassion begins in the home, and in our hearts, it can be distorted by both less than fully developed feelings, or by cognitive misunderstandings.

Of course, the main response to this objection has already been given, in Gruen's account of "empathic inaccuracy" and in my (the Dalai Lama's) universal compassion. These are not just subjective responses to suffering. Real compassion does not involve pity. Pity is the unmodulated attitude of looking down on helpless others. Universal compassion, in contrast, sees the causes of suffering, in both self and other, and it therefore prepared to act to relieve suffering. Stated in my terms, just as the epistemic objection doesn't understand the distinction between emotions and feelings about emotion, so the moral objection doesn't understand the difference between empathy and compassion.

However, the moral objection might cause us to make one concession. Since empathy, especially, can be challenged both by bias and by scale, we sometimes need a more impartial view of those we love. And in a globalized world we cannot always depend on common intuitions about what counts. Just and fair legal institutions play this role on a larger scale of offering

1977. The feminist-vegetarian Bloodroot Collective is organized and creates their vegetarian restaurant in Bridgeport, CT.

basic guarantees of equal participation. At their best, they can correct our biases. Justice is not, in principle at least, at odds with compassion. They are complementary perspectives. The only kind of justice worth having is justice administered with compassion.

Food as a compassionate practice

Who is it, then, that engages in a moral practice? We know that for political liberalism it is a disengaged self, the strongly individuated self that can be the owner of individual rights. In contrast, the image that emerges from the above characterization of compassion is markedly different. Here we encounter a body-mind thoroughly engaged in a social world from the very beginning. In the terms I've used, it begins in evolution in the F5 part of the brain that has evolved to make social meaning possible. A significant finding of this sort of research is that our moral lives do not start in a solitary, Cartesian inner space that later becomes social. Empathy is constituted by its sociability. Its most basic function is to make the world of self and other meaningful. Our social selves are basic.

In contrast to this picture of moral functioning, it is not surprising that politically liberal animal rights philosophers seem contorted, as if performing a feat in intellectual gymnastics, when they "extend" human moral standing to non-human others. In one famous account, some non-human others are "subjects of a life" (Regan and Singer 1989) if they are sufficiently like human beings in their ability to be self-conscious, rational beings. Rights are directed at humans, or, more precisely, persons (rational beings, including humans). Others get moral standing by analogy to us.

Far from requiring moral extensionism, the compassion approach starts from a common source. We are far from alone in being able to experience empathy, and as Metzinger's examples show, other species very likely show extraordinary kinds of empathy that are not very well developed in humans—for example, the shapes of flocks of birds in flight. Or, in Damasio's terms, single-cell organisms exhibit biological value, the drive for homeostasis. At a much more complex level, "A number of mammals" have what he calls an autobiographical self, "namely wolves, our ape cousins, marine mammals and elephants, cats, and, of course, that off-the-scale species called the domestic dog" (26). The compassion approach begins with thorough engagement in a common physical and social world. Compassion grows out of insight into the connectedness of self and others. It is not, originally, directed exclusively toward humans, but toward all the co-inhabitants of the social world. From the very beginning, for example, the Buddhist moral universe has been "all sentient beings," not just "persons." In short, the issue of "other beings" is a core element in any richly developed ethics of compassion, not simply an add-on.

What would this picture of an ethic of compassion look like when applied to our engagement with non-human others, and to food? Well, I suppose much of this is obvious. Compassion is a commitment to all sentient beings, not just humans. It begins in our deep-seated emotional connections to other beings. It has no need, therefore, to privilege rational personhood, and extend the boundaries of the moral to other beings if they are like us.[8]

Food is not a natural category. Much of what is edible and nutritious is not usually counted as food—other human beings, for example. On the contrary, much of what counts as food in the American diet is barely edible and far from nutritious.

If Carol Adams's work has shown us anything, it is that what we count as food is a deeply contested category. Gender, race, class, caste are all deeply imbedded. And so, our food practice becomes an avenue through which non-violence can be realized, where wisdom and feeling come to coincide. I have said for many years (Curtin 1991) that I think it's best if we view this as a direction rather than a place.

There are many vectors now, not only a more compassionate, therefore less violent eating practice, but environmental issues: how far does our food need to travel? How was it produced? Was there economic fairness for the producers? Is quality food, given the odds against it in our economic system, too expensive for those who are not among the economic elite? And things change. At one time a diet that included fish may have seemed somehow less violent and more sustainable. It's hard to sustain that position now.

So, a food practice calls for practical reason, constantly balancing competing factors. It calls for compassion, most of all for the direct victims of our dominant food practices, but also those who are exploited by the contemporary industrial food system. It should also evoke some compassion for ourselves as we work through this complex set of demands to achieve a life whose flourishing does not diminish the flourishing of others. And, if the Dalai Lama's challenging ideas about universal compassion make sense, we need to practice compassion even for those who support the industrial food system, not in spite of the slaughter of billions of innocent beings, but because of it. As the philosopher Baruch Spinoza observed, "Hatred is increased by reciprocal hatred, and may on the other hand be destroyed by love" (Spinoza and Morgan 2006, 83).

Finally, an overriding metaphor. The opponents of compassion ultimately see abstract rationality as providing rules by which we can distinguish good

1977. Dr. Daphne Sheldrick establishes the David Sheldrick Wildlife Trust in Kenya to protect and rehabilitate elephants.

from evil. If compassion starts from such a different place—the transformation of emotions into feelings—is its goal also different? Can it be that the friends of rights and the friends of compassion are talking about two different things when they use the word "ethics"? I suspect so.

I think a compassionate food ethic does grow from a different source, and here again I cite Śāntideva. The basic Buddhist vow of compassion is: "I am medicine for the sick. May I be both the doctor and their nurse, until the sickness does not recur" (Śāntideva, Crosby, and Skilton 2008, 20). The Dalai Lama was invoking qualities traditionally associated with both the doctor and the nurse in defining universal compassion. Those who vow to engage the world compassionately need both a deep understanding of the causes of suffering, and yet also to engage this patient, this particular history, and this family ... Really seeing what's best, all things considered.

Compassion is a form of practical therapy. The model for an ethic of compassion is health, not cognitive correctness.

Acknowledgments

This chapter was written with the support of a Fulbright-Nehru Fellowship and the United States-India Educational Foundation as well as a sabbatical from Gustavus Adolphus College for 2012–13.

Notes

1   See Olmstead vs. United States 277 U.S. 438 (1928).
2   When His Holiness, the Dalai Lama, argues that the Chinese are violating Tibetan human rights when they torture political prisoners, this has a universal ring to it that may be more effective than appealing for compassion. I would note that His Holiness often does appeal to compassion and to rights. This was also a central feature of Gandhi's moral claims against the British.
3   In general, the topic of neuroplasticity is fascinating, and important to a complete account of moral development.
4   Much of the original research in this area was conducted by opening up the brains of macaque monkeys. And research on sociability is often conducted on non-human primates who are deprived of social interactions. There are obvious moral problems with such approaches.
5   I have spoken to many Tibetan refugees who learned only many years later from newer refugees that their entire families had died.

6   See Dalai Lama, "Compassion and the Individual."

7   It might be objected that the comparison of the Dalai Lama with abuse survivors amounts to the inappropriate expectation that women be moral heroes. There is a difference between expectation and recognition. Universal compassion is necessary precisely in the most demanding cases where empathy is impossible or inappropriate. The problem here lies more with our preconceptions about what a hero is than with the actual achievements of women who find a way to health after suffering abuse.

8   I would also point out that, while neuroscientists are now very comfortable with the idea that mind extends deep into nature—earthworms display mental activity—there is a clear divide with plants. Plants do not have neurons. I do not mean to suggest that an ethic of compassion is irrelevant to an environmental ethic, only that an environmental ethic raises different issues, which are outside the scope of this paper.

1978. Susan Griffin's *Woman and Nature: The Roaring Inside Her* published.

Joy
Deborah Slicer

3

We don't eat those with whom we play, joke, laugh. This isn't an empirical claim. In fact, some people do eat those with whom they play, joke and laugh. But many of us find it odd, even incomprehensible, a kind of category mistake, or worse. And when I say we don't eat those "with whom" we laugh, I don't mean laugh at (derisively) or laugh about (fondly), but laugh with, share something humorous.[1] While I do think other animals have very significant moral claims, I don't want to go there through the usual inclusivity and consistency arguments (inclusion of cognitively "sub-normal" humans in the moral universe requires we include cognitively similar non-human animals), however much I respect the huge inroads those approaches have made into academia and popular discussion of animals' moral and legal standing. And I grant that these arguments have accomplished this because they are largely sound. For a long time I have been more attracted to the work of Cora Diamond and more recently to the novelist J. M. Coetzee's *The Lives of Animals* (1999). Diamond, because of her Wittgensteinian insistence that we take up ethical questions within the socially convoluted and epistemologically vexing forms of life that give rise to them, because of her insistence on literature and life as the best moral teachers, and because as early as 1978 she began working on the moral significance of other animals as "fellow creatures," not just bearers of sentience or as subjects of a life, but as those with whom we share a certain "boat" and seek out as company, a theme that a number of other writers since have riffed without giving her due credit (Diamond 1991a, 329). And Coetzee, because his Elizabeth is so thoroughly the utter astonishment, bitter frustration, rage,

1979. Founding organization of the Animal Legal Defense Fund was started.

alienation, loneliness, and despair that many of us who are long-time vegans and activists frequently feel among families like Costello's, among students and colleagues like Costello's, at dinner parties much like the one in honor of Costello, when we must shop for food and shoes and handbags, just like Costello. My concern for Elizabeth is that she seems humorless, and laughter is joy juice, an affirmation of life. This is one of those great "difficulties of reality," to use Cora Diamond's phrase, that joy along with profound, senseless suffering exist in equal measure as givens of our thrown condition, and, that whether taken separately or standing side-by-side, they are both equally freighted moral spaces. A certain kind of moral courage is required to be with that, particularly with their side-by-sideness, not just to intellectually acknowledge it, but to experience that tension in the deepest existential sense, and to keep one's head attached and one's moral bearings.

Story telling

Certain writers say that philosophy must incorporate stories/narrative into a responsible representation of and response to moral life. Iris Murdoch (2001), Bernard Williams (1985), Cora Diamond (1991), Martha Nussbaum (1992), Stanley Cavell (2008), various feminists and ecofeminists, including Marti Kheel (2008), are among them, and their work is familiar by now. Like Aristotle, they show how literature is instructive because it represents something authentic in its difficulty and because it requires we participate—that we practice attention, emotional sensitivity and sensitivity to detail, deliberation, choice, and judgment along with the narrative's characters. And of course those characters model all this for us. A bit more specifically, Nussbaum and Diamond, for example, note the way traditional philosophical treatments of moral life flatten and distort complex and emotionally charged moral situations into Kantian or utilitarian riddles that can be handily resolved by applying over-simple, abstract principles to them, by treating moral dilemmas as math problems with people. What is lost, among other things, is an understanding of the difficulty of choosing well, to paraphrase Nussbaum, within a context of various and competing particulars all vying for attention. And then there's the difficulty of choosing and relinquishing competing, noncommensurable goods: erotic love or filial duty, or a truth that threatens friendship, for example. Robert Solomon (1976), Nussbaum (1992), and Diamond (1991, 2008), among others, also recognize a place for emotional intelligence in moral life. By "intelligence," I mean that the emotions have some legitimate sway over the will and that they have important cognitive functions. All this raises issues of what the philosopher should be doing. Should we be reading stories, become better literary critics? Should we teach stories alongside of or instead of "philosophy"? Should we write stories? Diamond and some ecofeminists, such as Marti Kheel (2008), Chris Cuomo and Lori Gruen (1998), and Greta Gaard (2007), stress the importance of stories because they help us see and feel that and how animals are subjects. As Kheel puts it: "Stories, such as Babe the pig's and Emily the cow's, help people recognize that the lives of other animals follow strong lines, representing a subjective identity. One way in which we can come to appreciate their subjectivity, therefore, is through telling their stories as best we can" (Kheel 2008, 249). Cuomo and Gruen advocate "transgressive moral orientations" that bring the background into the foreground, including non-human animals' subjectivities into the foreground of our moral attention. Stories are transgressive because they contain "ins" that disrupt our usual ways of perceiving and feeling (or not feeling) and make possible empathy along with real friendships with animals—a point I'll return to later (Cuomo and Gruen 1998, 133–4). The following stories inspired this piece. They also help me work some of the more difficult points I'm trying to make about the moral significance of playing and laughing, of creating joy with another. They don't just illustrate the points. They do their own work and have their own depth, I like to think. And ultimately I hope they're transgressive, "ins."

Humor

Asa, a 21-year-old, 16-hand, black quarter horse, is most in his element with groups of younger geldings who probably see him as the cool, middle-aged guy who shows them how to negotiate a variety of social situations—which crankier horses mean business and which bluff, how to move a guy off his hay pile without expending excessive energy or getting hurt, how to kick the ranch manager's pastern-nipping Shepard while not crushing the little pest, how hard to run the pants off a new equine until or if he assumes the bottom of the pecking order, which two-legs are going to give you real grief if you don't take orders and which, despite their predatory stature and bullying behavior, you can tell to take a flying fuck by simply playing deaf or, if you've really got some chutzpah, delivering a premonitory kick. He's the older guy who, despite an arthritic lameness in his right front knee and left hock, is good for an impromptu, screaming rip around a field after wild turkeys or deer or a falling leaf or puff of wind or some seismic shift—another

1979. Elizabeth Dodson Gray's *Green Paradise Lost* published, one of the first books to theorize a relationship between heterosexism and the oppression of nature in patriarchal cultures.

herd running a field a mile distant. Asa and I have been together for nearly ten years. He's one of the funniest guys I know.

A word about horse humor. Play and humor are related expressions of élan vital, pure being, our "life force," to use Henri Bergson's (1974, 1998) and Susanne Langer's (1965) expressions. Laughter, she says, is the "culmination of feeling—like the crest of a wave of felt vitality" (132). Horses have élan in spades. They have devious senses of humor, and they're wicked practical jokers—butt biters, tail pullers, gate openers, bucket-, brush-, and hat-snatchers, stealthy farters, among other things.

Early in our relationship I practiced Pat Parelli's natural horsemanship, a series of what Parelli calls "games" that improve communication between horse and human, help establish the human as herd leader/teacher, and keep everybody safe. I've since had some misgivings about whether the natural horsemanship movement is as noncoercive, as egalitarian, attentive to the equine other, and as spontaneous as I want my friendships with horses to be. That said, these schools do stress understanding the horse's mind, clear and consistent communication, and low-key mutual fun. This particular program was helpful to us during our first year together when Asa sized me up as a harmless novice and object of comic torture.

So I spent many hours watching Parelli DVDs and reading Parelli handbooks, taking copious notes, condensing pages of notes into field instructions that I'd reference as I "played" with Asa, who, haltered, stood at the end of a 15-foot lead line, patiently, usually. One afternoon as I studied my notes, which I'd written on the back of an envelope, a pair of thick, black horse lips very gently took possession of the top edge of the paper. Horses have bilateral vision; they can see two different images spontaneously. But they can also converge their focus more directly on a single image in front of them, so long as they have a little distance from the object. Asa found a distance and adjusted his head down a bit to focus on this very funny joke. Many horses' faces are exquisitely expressive, though I have often seen in horses the same gaze that John Berger describes in zoo animals—a "gaze that flickers and moves on … They look sideways. They look blindly beyond. They scan mechanically" (cited in Coetzee 1999, 34). But Asa's black eyes were soft, round, directed and energetic, an energy that, as hard as I try not to, I always associate with a jolly Santa, or, what's more to my liking, with my grandfather, who had a similarly dry, teasing sense of humor. Ears up, chin loose as a wattle, his eyebrows twitched across his forehead like heat lightning. At first I protested for real, tugged the paper, and then tried backing him, knowing that in horse-land whoever moves the other's feet is dominant. All for naught. The lips trembled the paper and held fast. My path to "love, language, and leadership," the Parelli guarantee, was clinched between my horse's front teeth. How absurd was that? And while tugging at my new social "contract" with my equine partner, I began to laugh. Many horses associate human laughter, that particular vocalization, with good things—petting, cookies and carrots, and just a general good feeling; it's a vocal cue to cock a leg, yawn, lip-lick, let your mane down. Asa cued to all that, and I suspect he may have also heard my laughter as uptake, that I was getting the joke. At which point he began very slowly, and without breaking eye contact, lipping the entire page into his mouth. Once he'd had it all he chewed it leisurely, like an old man with his plug. Then casually spat the soggy wad into the dust. R. H. Smythe in a wonderful little book called *The Mind of the Horse* comments that "some horses possess a certain sense of humor and are prepared also 'to act daft' occasionally in order to pander to the whims of humanity" (Smythe 1965, 86). He doesn't elaborate. I've found precious little else on other animals' senses of humor, apart from some bits on laughter by Marc Bekoff, short pieces by dog trainers, and a little snippet titled "The Rat Who Laughed."[2]

Theories of humor consistently highlight how things we find humorous subvert our expectations, whether physical or linguistic or social, without seriously painful or offensive consequences. We don't laugh at something that we truly consider true, beautiful, or good, according to Robert W. Corrigan, though often comedy walks a dangerously fine line (Corrigan 1965, 6). A number of writers, including Bergson and Hazelit, stress incongruity and the subversion of expectations as potentially humorous—an umbrella the wind turns inside out or a clown's or mime's exaggerated, almost mechanized, limbs and facial expressions. Al Capp's theory that we delight in "man's inhumanity to man" reverberates with Langer's and Freud's claim that we find comedic delight in our sudden, unexpected superiority to some pathetic fop, brilliantly evoked by Charlie Chaplin, Capp argues. Susanne Langer's discussion of the old folk character "Punch," who carries out every repressed impulse by "force and speed of action—throws the baby out the window, beats the policeman, spears the devil with a pitchfork," reminds me of Eric Bentley's claim that farce is violence without fear of serious consequences (134). Aristotle, in his *Poetics*, considered comedy much inferior to tragedy and a species of the ugly, the ludicrous, which isn't painful and causes no harm. Freud claims that "wit is an escape from authority, nonsense an escape from reason" (in which case witty Asa was full of nonsense).[3]

This is a very thin sampling of theories of comedy, and of humor, which is different, and a flimsy treatment of that sampling. What's most relevant here is that almost invariably when other animals are mentioned theorists deny that they have senses of humor. Largely they follow Langer's logic: other animals are not self-conscious, they simply pass through the natural

1979. Elizabeth Fisher's *Woman's Creation: Sexual Evolution and the Shaping of Society* published.

succession of "individualized existence"; theirs is an instinctive struggle for survival, while the human animal "ponders its uniqueness, its brevity and limitations, the life impulses that make it, and the fact that in the end the organic unity will be broken, the self will disintegrate and be no more" (125). Very young children lack senses of humor too, even though they laugh, a much more elementary thing than humor, she says (132). Because children do not yet have a sense of a "self" that they must willfully negotiate through the social and natural worlds and that are battered by chance and fate, and because they have no conceptions of their own vulnerability and finitude, children have no "sense of life," and thus cannot take pleasure in bizarre, absurd, improbable, stumbling, frivolous attempts to subvert and/or cope. One requires a "semantically enlarged horizon" for all that (124).

But suppose Langer's wrong. The facts. The arguments. *Her* sense of life. Suppose this time we give credence to anecdotal evidence. Trust the gut, love's knowledge. Suppose Asa and I joke. Play. Suppose we seek them out as fellow creatures because we laugh together. Suppose our boat rocks with laughter. What kind of moral space have we entered?

Play

Animals play. What exactly is "play"? Who plays? Why do animals play? When do they play? What functions, if any, does play serve? What are some of the communicative and cognitive aspects of play? And what are the social costs of not playing? Marc Bekoff and Colin Allen distinguish between functional and nonfunctional definitions of play. A functional characterization relates play to motor, social, and cognitive development. In 1998 Bekoff and Allen were leery of functional accounts because there were so few studies to support those theories (99). But in 2009 in *Wild Justice*, Bekoff and Jessica Pierce develop a functionalist-sounding theory of play as training for moral life, which they define as "a suite of other-regarding behaviors falling into three rough clusters of cooperation, empathy, and justice" (138). "Morality is a spectrum of behaviors that share a common feature of concern about the welfare of others" (138). Drawing on the cognitive literature they establish certain "threshold conditions" for at least mammals as moral animals:

> [A] certain level of complexity in social organization, including established norms of behavior to which attach strong emotional and cognitive cues about right and wrong; a certain level of neural complexity that serves as a foundation for moral emotions and for decision making based on perceptions about the past and the future; relatively advanced cognitive

capacities (a good memory, for example); and a high degree of behavioral flexibility. (13)

Play is important to acquiring certain moral skills, such as developing basic social skills, bonding, learning social norms, reciprocity, respect for rules and others' social space, recognition and assessment of others' intentions, responsiveness to surprise and novelty. Playing fair, following the rules, is important; cheaters stop the game and are often chased out. Among other things, they've violated trust and may well be shunned. Trust, in particular, Bekoff and Pierce suggest is necessary for group cohesion and stability. In play we often neutralize inequalities and promote egalitarianism. Self-handicapping is a good example. Other animals, just like us, may do things they might not do outside of play—role reversal, for example— something that may even compromise their wellbeing in nonplay situations. And play involves justice, a natural sentiment, a feeling, the authors claim, and not necessarily an abstract set of principles.

Nonfunctional characterizations define play as "all motor activity performed postnatally that *appears* [authors' emphasis] to be purposeless, in which motor patterns from other contexts may often be used in modified forms and altered temporal sequencing. If the activity is directed toward another living being it is called *social play*. This definition centers on the structure of play sequences—what animals do when they play—and not on possible functions of play" (Bekoff and Allen, 99). Nonfunctional definitions are problematic because they apply to nonplay behavior such as excessive grooming and repetitive pacing by caged animals. So Bekoff and Allen find a third way, a sort of behaviorist account:

> Our view is that the study of play ought to start with examples of behaviors which superficially appear to form a single category—those that would be initially agreed upon as a play—and look for similarities among these examples. If similarities are found, *then* we can ask whether they provide a basis for useful generalizations. We therefore propose to proceed on the basis of an intuitive understanding of play, guided to some extent by Bekoff and Byers' attempt to define it, but without the view that this or any other currently available definition strictly includes or excludes any specifics behaviors from the category of play. (100)[4]

Horses play as often as dogs and, curiously, they play many of the same games as dogs, whose ancestors were once among horses' natural enemies. Horses play chasing games, neck-biting games, tug-of-war, and take-away games, among others. Asa and I play games, though I don't rough-house with my 1,200-pound friend. We do play take-away. Often the stolen object

1980. Northern Animal Liberation League active in the north of England, with a motto of "Over the fence when they least expect it."
1980. People for the Ethical Treatment of Animals founded.

is my hat, sometimes a brush, sometimes a bucket. Sometimes we play tug-of-war with these items or with a lead rope. Occasionally we play a chasing game, which is simply a chase-me-if-you-want-to-halter-me sort of thing. We did this a few times when Asa wanted to establish dominance in our relationship, long ago, and then it wasn't so funny. Now I know it's a game because his body language is entirely different, and anymore he nearly always trots up to me at the gate, where I wait, resting on a boot toe, just another relaxed horse. Another game we play and that he's recently started initiating is the transition game. Most riders, including myself, like very quiet transitions up from a halt to a walk, to a trot, and then to a canter, and then back down. Asa and I transition on the ground in his 10-acre bedroom. Walk, trot, halt, back up, trot, etc. We mix it up and try to keep an element of surprise going—a screeching halt or we jump obstacles like branches or small ditches. He handicaps himself so we can trot or canter side-by-side. The first time he initiated the game I was walking out in front toward a gate, when he trotted up beside me, nearly trotted in place so as not to out-distance me, and something in his facial expression, mostly in those effervescent eyes, said: "P-L-A-Y?" As I said, he handicaps himself, a lot. He's kind and hangs with me. Rules: One guy doesn't run off and leave the other, we don't play this game when he's haltered, we don't ask each other to do things that are dangerous or impossible, like run through an acre of prairie dog holes or very close to a fence line or jump over gates, and if the other horses in the field join us, we stay focused. These aren't just my rules. They've evolved between us, over time, and as we've paid attention to one another.

My friend Jeff Hudson says that horses have to feel they're on equal or near-equal terms with a person before they'll horse around with you, joke or play. Barbara Smuts writes about equality—reciprocity, freedom, mutual dependence and respect—as the bases for friendship with another animal. Recognizing and allowing another to be a social subject and an idiosyncratic individual requires, among other things, we give up "control over them and [importantly] how they relate to us" (Smuts 2008, 118). I rarely hear anyone in the horse world, even those who practice natural horsemanship, talk about equality, though many talk about the importance of getting inside the horse's mind, which is at least some acknowledgment of their subjectivity. Mostly we get inside the horse's mind though in order to more effectively control them. I handily concede that one has to use more than usual caution when playing with a 1,200-pound animal whose flight response might be triggered by a waving branch or a garden hose or an odd-shaped rock along the trail and who isn't always cognizant of my personal space or their respective physical advantage. But for me this means that developing trust is more, not less, relevant than in my other relationships. Play and joking around with a very large animal can be dangerous. Nonetheless if I want this in my friendship with horses, I have to give up significant control and to trust, at minimum, our communication, that we'll follow the rules, that we'll take care of each other when one of us innovates, and, importantly, I think, when we play and joke we trust each other with a precious gift we're sharing, our own joie de vivre, that vital force, which so marks our kinship with all other living individuals and with larger cycles and creative processes from evolution to poetry, the joy that other animals, from steers to rats, unconditionally give themselves over to in play. Everything I've just mentioned here has moral traction. Cognitively, to have a sense of humor it's necessary to subvert and to recognize subverted social, physical, and communicative expectations and sometimes to make fine-grained distinctions between subversions that are very inappropriate because they're harmful—painful or offensive—and subversions that are benign, sometimes silly, and give us joy. And play involves, among other social, cognitive, and emotive capacities, the recognition of rules, respect for rules, trust, an appreciation for novelty, and a sense of fairness. I think that other animals have these capacities, and that they have senses of humor and that they play. And most ethicists, short of perhaps the doctrinaire Kantian, will acknowledge that these things matter very much, morally. What most interests me in the following is what it means to mark our kinship, our status as fellow creatures, as Diamond might put it, by sharing joy as tricksters and playmates.

Two points, briefly, before I close this section. First, I didn't intend to convince skeptics that animals do in fact play. The literature on play is extensive and worth reading. Better yet, since gross contact is the soundest proof and the most successful moral "in," go there first. Second, humor and play are cousins, but they aren't the same. Play is fun, but it's not always funny. Funny isn't always a game.

Kip

Within the last year I helped retrieve several horses who passed through the auction at the Missoula Livestock Exchange, all of them destined for slaughter. The first, an old friend, sold to a kill buyer by his human companion and former friend of mine, is Kip. Because Kip was in good weight and in good general health he was "fast tracked" to Canada from Montana, which means he should have been "shot and hung," as the industry puts it, within three days of sale. An hour after the sale he was loaded as hip number "138" onto a double-decker livestock trailer, which, even though such trailers are

1980. Carolyn Merchant's *The Death of Nature: Women, Ecology, and the Scientific Revolution* published.
1980. The Bloodroot Collective's *The Political Palate: A Feminist Vegetarian Cookbook* published.

unsafe for horses, is a common means of transporting them to the slaughter plant in Alberta, a six-hour trip. Minutes prior to departure, a backcountry outfitter pulled him off that truck, trading two of his horses, "all broken down at the knees," as he later told me. After that Kip was in limbo for five weeks until I found him in Ennis, Montana. This all happened in the back lot of the stockyard, and I had no idea that Kip was alive during the five weeks I searched for him, having been encouraged by two friends to keep looking and by a nagging sense of my own that I could and needed to find him. Whether that sense was my inability to get my mind around a "friend" being someone else's "meat," an insurmountable "difficulty of reality," to borrow Cora Diamond's idea, or some kind of energetic connection to him that's as difficult to explain, I don't know (Diamond 2008, 45). During that five-week search I talked with kill buyers, people at the Canadian slaughter plant, feed lot workers, brand inspectors, rescue workers, stockyard employees, and I now have a fair sense of how this system works, a system that's both legal and clandestine, very poorly regulated, and that, for me, was accented by moments of disorienting kindness.

During the six years I knew Kip prior to his sale he was a complete gentleman on the ground, personable, kind, and, mostly mustang, he had a mustang's mind—exceptionally intelligent, agile, quick. People who knew horses were inevitably attracted to him. Kip also made it abundantly clear that he would not be ridden. At eleven years old he still refused to carry a rider, whether because of past abuse, because of his unusual conformation that sometimes caused him pain, because of ill-fitting saddles, nobody knows. When his person insisted on riding, Kip gave it an honest try, then decided his original fears were grounded. He bucked, hard; she got hurt. And what good is a horse you can't ride? Or so it goes with many horsemen and horsewomen. It's not an uncommon story. Kip is now my companion, and I have no expectations of him. A new trust has to begin there.

Being

In "Eating Meat and Eating People," Cora Diamond writes about the distinction we make between "pets" and animals people eat. Once we understand how the concept of "pet" functions, we understand that we don't eat them. A pet "is given a name, is let into our houses [horses excepted] and may be spoken to in ways in which we do not normally speak to cows or squirrels" (Diamond 1991a, 378). Our concept of other animals entails their being fair game as food, clothing, research and testing subjects, and as entertainment. In the case of the not-pets, perhaps we stress the differences between "them" and "us," while stressing the similarities between pets and ourselves. These concepts are social constructions and as such they are somewhat fluid. Cultivating the empathetic imagination, as ecofeminist and other writers, such as Diamond suggest, is key to recognizing similarities, especially cognitive and social ones, and insomuch as the heart is involved, these similarities, mostly these possibilities for deep social relations, can take on moral weight. The simple fact of being alive, of exercising a telos, as Paul Taylor argues in his *Respect for Nature*, as well as differences, differences that are mysterious, awesome, humbling, can take on moral significance too of course (Taylor 1986). But the cognitive and social similarities are easiest for us to notice because we are profoundly problem-solving and social creatures ourselves.

Horses are exceptional; the conceptual boundaries are much looser with horses and other equines. Horses are both pets and meat. While it's true that the majority of horses who end up at auctions and go to slaughter are there because they lost their economic or recreational value for a human being, those animals also have names (but become "hip numbers" on the auction floor), were socially educated/trained by human beings, were likely groomed regularly, have recognizable "horsonalities" (to use Parelli's term), were praised and petted, and encouraged to trust, all the sorts of interactions typical of our relations to so-called pets. And the same individual who had these social interactions with the horse may very well send the horse to slaughter in a legal and social context that, even though the majority of Americans disapprove, enables the deaths of approximately 130,000 US horses each year. The conceptual distinctions we make between "them" and "us," those who are at the table and those who are on the table, to paraphrase Diamond (1991a), are slipperier than usual in the horse's case, and this conceptual instability is reflected, among other places, in our ambiguous relationships with their bodies, with their "fullness of being," to use Coetzee's character Elizabeth Costello's term (1999, 33), which our many ingenious and largely successful mechanical "aids," from curb bits to hobbles, are designed to lock up. I'll return to this shortly.

Both Diamond and Costello consider acts that are beyond the pale. Diamond, who comments on why it's a kind of mistake to say it's wrong to raise people for meat, to salvage the dead for their organs, supper, or the compost heap, points out that to say these things are wrong isn't too weak, "but in the wrong dimension" (Diamond 1991a, 323). Costello's many references to the Nazi extermination camps could be her attempt to make a similar point about something beyond the pale, and, for better or for worse (Coetzee lets us decide), she draws analogies with the mass slaughter of animals.

1981. Sally Gearhart gives a talk on interconnected oppressions at the rally on World Day for Laboratory Animals.

Something beyond the pale, of another dimension, is related to its being a difficulty of reality, to use Diamond's term again. My attempts to understand Kip's fate and the equine slaughter industry resonate with this kind of difficulty. To be clearer, here's Diamond. This difficulty "is experience of the mind's not being able to encompass something which it encounters" (Diamond 2008, 44). Later she says these are "experiences in which we take something in reality to be resistant to our thinking it, or possibly to be painful in its inexplicability, difficult in that way, or perhaps awesome or astonishing in its inexplicability. *We take things so*. And the things we take so may simply not, to others, present the kind of difficulty, of being hard or impossible or agonizing, to get one's mind around" (45–6).

While some find enigmatic Costello's comment that she's a vegetarian in order to save her soul, I think it makes sense given her profound existential disorientation, feelings of isolation, and her frustrated attempts to explain herself to an unfriendly audience of highly educated, socially graceful, kindly-seeming people at Appleton College. She even says to her son: "It's that I no longer know where I am" (Coetzee 1999, 69).

> I seem to move around perfectly easily among people, to have perfectly normal relations with them. Is it possible, I ask myself, that all of them are participants in a crime of stupefying proportions? Am I fantasizing it all? I must be mad! The very people I suspect produce the evidence, exhibit it, offer it to me. Corpses. Fragments of corpses that they have bought for money…Yet I'm not dreaming. I look into your eyes, into Norma's, into the children's, and I see only kindness, human-kindness. Calm down, I tell myself, you are making a mountain out of a molehill. This is life. Everyone else comes to terms with it, why can't you! *Why can't you?* (69)

And her son John, a mixed salad of repressed, raw emotions, who throughout the book mentally notes her old flesh, slumped shoulders, white hair, and old-womanish cold cream smell, whispers in her ear: "There, there. There, there. It will soon be over" (69).

This exchange takes place in the car on the way to the airport after Elizabeth finishes at Appleton. As the novel's closing line, it's brilliantly ambiguous. Is he referring to Elizabeth's departure from Appleton? Or to the end of his own ordeal as his mother's host? Or is he trying to soothe her by promising some kind of (unlikely) pro-animal revolution? Is he referring to her death? Given that the comment immediately follows his mentally noting, yet again, her aged body, it's most likely he's referring to her dying. Much like the way we ontologize other animals, her son frequently comments on Elizabeth's body, equates her with her mortal corporality, and in that way (among others) diminishes, or outright rejects, her message, and ultimately her subjectivity. I wonder if a woman with more vital energy, which is not the same as youth, with what Elizabeth herself calls "fullness of being"—joy, for example—might be harder to "kill off," to ontologize with death. Coetzee portrays Elizabeth as a victim, as an existentially "wounded animal," to use her words, defending other victims. And while I agree with Costello's points about the existential and mortal damages of our relationships with other animals—our victimization of them and our subsequent own self-destructiveness—and I recognize this as the most prevalent type case against our treatment of other animals (and I've made the case myself in essays and as an activist), here I offer a different notion of subjectivity and relationship that, without taking a familiar "like-us" approach to animal relations, empha-sizes our shared condition and our ability to respond to it through humor and play. Cora Diamond's notion of animals as "fellow creatures" is relevant and in the preceding I've worked that idea by exploring our mutual capacity for joy. Costello grasps the idea too, but she only takes it so far.

Excoriating Descartes, Costello says:

> To thinking, cogitation, I oppose fullness, embodiedness, the sensation of being—not a consciousness of yourself as a kind of ghostly reasoning machine thinking thoughts, but on the contrary the sensation—a heavily affective sensation—of being a body with limbs that have extension in space, of being alive to the world. (33)

"Being alive *to* the world," "a heavily affective sensation," the sensation of being a body with limbs that have extension in space, to be full of *being*. And she says that "one name for the experience of fullness is *joy*" (33). Play, and especially the kind of physical, versus intellectual, play that young children and other animals so love, and much humor, laughter that crests the wave of "felt vitality," to cite Langer again, are manifestations of this sensation.

Costello says that when we are "full" we're alive "to" the world, not simply alive "in" the world.[5] So much hinges on that preposition. I want to hang with this phrase long enough to tease out some of its less obvious implications for joy. Even in our profane world, ours is a fallen state; we are creatures whose emotional and moral constitutions are as malleable and durable as tinfoil in a world so resistant to even our best efforts that it's virtually impossible to avoid guilt, shame and estrangement, and grief, grief as inevitable as death. Or better yet, as Herman Melville's Ishmael puts it, "there are certain queer times and occasions in the mixed affair we call life when a man takes this whole universe for a vast practical joke, though the wit thereof he dimly discerns, and more than suspects that the joke is at nobody's expense but his own" (214). The universe certainly had the last laugh on the Pequod. Ishmael's job was murder, as was Melville's for three years in the

South Pacific. And yet their voices are as "alive to" as any I've ever heard. Ishmael:

> Days, weeks passed, and under easy sail, the ivory Pequod had slowly swept across four several cruising-grounds...It was while gliding through [the Carrol Ground, an unstaked, watery locality, southerly from St. Helena] that one serene and moonlight night, when all the waves rolled by like scrolls of silver; and, by their soft, suffusing seethings, made what seemed a silvery silence, not a solitude: on such a silent night a silvery jet was seen far in advance of the white bubbles at the bow. Lit up by the moon, it looked celestial; seemed some plumed and glittering god uprising from the sea...[And] some days after, lo! at the same silent hour, it was again announced: again it was descried by all; but upon making sail to overtake it, once more it disappeared as if it had never been. And so it served us night after night, till no one heeded it but to wonder at it. Mysteriously jetted into the clear moonlight, or starlight, as the case might be; disappearing again for one whole day, or two days, or three; and somehow seeming at every distinct repetition to be advancing still further and further in our van, this solitary jet seemed forever alluring us on. (Melville 1964, 219–20)

These "spirit spouts" inspire both awe and blood lust in the crew, solidarity with the universe and estrangement from it, a "quivering" pleasure and a terrible premonition. Costello doesn't go here, but I say that when human *being* is alive to the world it is alive to all these things, that is to paradox, mystery. In any case, maybe because Costello, alive to the world, holds that paradox too long and is nearly mad, or because she's constitutionally gloomy, or because she lacks a certain courage and emotional fortitude, she is joyless. And what I've been getting at is this: When we sail our little boat over the dark, open water, its "scrolls of silver," "suffusing seethings," and "soft silence," take a butt-biter with you, play a flatulent game of chase. Because joy is part of the cosmic deal too.

Courage

In a remarkable short essay by the British playwright Christopher Fry he says that most of his comedies began as tragedies, that if the characters were not "qualified for tragedy there would be no comedy," and to some extent, he says, "I have to cross the one before I can light on the other...A bridge has to be crossed, a thought has to be turned. Somehow the characters have to unmortify themselves: to affirm life and assimilate death and preserve joy" (Fry 1965, 17).

I said earlier that I worry about Elizabeth Costello because she seems humorless and later I said she was joyless. On the one hand, Elizabeth is very much alive to the world, a world in which the holocaust on animals, as she's describing it, exists alongside "kindness, human kindness," and even joy, a difficulty of reality as stubbornly resistant as any. On the other hand, she has no fullness—she's wounded, traumatized, suffering a "profound disturbance of the soul" (Diamond 2008, 56), precisely and understandably because she's trying to hold this difficulty in her mind. I believe that even, especially even, had we not been unfathomably lucky last year, had Kip been "shot and hung" along with the 40 other horses from his sale, along with the unlucky two the outfitter traded for him, I believe I *should* be writing this same piece, a piece that acknowledges joy, affirms life. But I very much doubt I would be writing it. Like Costello I may well lack the emotional fortitude and a certain kind of courage that being alive to and that fullness requires. "I'm a poet and we tend to err on the side that life is more than it appears rather than less," Jim Harrison says (2011, 454). I hear this as a moral imperative.

One of my favorite characters in Coetzee's *The Lives of Animals* is an earnest, blunt British professor, who nails it when he says to Costello at dinner: "Therefore all this discussion of consciousness and whether animals have it is just a smokescreen. At bottom we protect our own kind. Thumbs up to human babies, thumbs down to veal calves. Don't you think so Mrs. Costello?" (45). Oh, yeah, I think so. I also think we're quite capable of sharing, even creating, joy with other animals, an affirmation of good and hope, a fire-orange torch sweeping the darkness that buoys our little Pequod. And we do not eat those with whom we breathe into that fire.

Notes

1   Without having any particular culture in mind, I do acknowledge that for some people other animals might be ontologized as both food and playmates. I do not think it's disrespectful to say that my worldview isn't ontologized that way. I'm even enough of a metaphysical realist to say that some ontologies, or at least some aspects of them, better correspond to the way things *really* are, and I'll even go so far as to say that I believe my worldview, *and* its moral fallout, is one of those. I don't think that saying this is *necessarily* disrespectful, e.g, imperalizing or in some other way arrogant. Some disagreements

among Christians or between Christians and secularists, for example, can be and often enough are like this. Exploring such differences and attempting to resolve them can be difficult, even crazy-making, as J. M. Coetzee's Elizabeth Costello shows us in *The Lives of Animals*, a novel I work with extensively in this paper. Reason is supposed to be the great equalizer and peacemaker of course. But Costello (and Coetzee in giving us his novels) shows us its flaws and limitations. Coetzee deliberately engages and at the same time argues that the imagination and the heart, along with reason, are moral operatives. And sometimes gross contact makes its own forceful case. When differences of ontology are worked over and sometimes resolved, I believe the "Leatherman" approach, a multi-tooled approach, does the job.

2   See Bekoff and Pierce *Wild Justice* (2009), 94 and 120; "The Little Dog Laughed—The Function and Form of Dog Play," by Cheryl S. Smith (2010); "Laughing Dogs," Patricia Simonet (2007); and "The Rat That Laughed," Jesse Bering (July 2012).

3   The authors cited in the paragraph all appear in Corrigan (1965).

4   Animals' intentions (beliefs and desires) is another difficulty for play theorists. Play, especially social play, involves pretense, or pretend. And pretense requires us to "read" each others' minds in order to distinguish play-fighting or play-fleeing, for example, from the real things. So we must be able to represent another's representation of reality, a second-order intentionality, according to Daniel Dennett. Bekoff and Allen put it this way, paraphrasing A. Rosenberg: "for animal a truly to be playing with b, it must be that a does d (the playful act) with the intention of b's recognizing that a is doing d not seriously but with other goals or aims" (101). For example, Asa knows that Deborah understands he's play-fleeing, not running from a real predator, that he's pretending. Dennett believes that some animals are capable of this. Critics argue they aren't and suggest they cue into context instead, body language or even pheromones.

5   My student Angela Hotaling pointed out this distinction in our seminar on animals and ethics in fall of 2011.

Participatory Epistemology, Sympathy, and Animal Ethics

Josephine Donovan

4

The assumption that animals are mechanical automatons who lack subjectivity and feelings, and whose communicative signs are merely biochemical reflexes, has long reigned in behaviorist animal sciences. Rooted in classical Cartesian/Newtonian science, which treats animals as soulless objects, this behaviorist view has legitimized animals' treatment as experimental material in laboratories, as commodities in animal husbandry and industrial agriculture, and as property under common law. This article is an attempt to explore and develop a participatory "subject-subject" conception as an alternative to the "subject-object" epistemology that defines classical science in order to undermine its legitimization of animal exploitation. For despite all its inadequacies as an explanation of living matter, Cartesian science—or what Thomas Nagel terms "reductive materialism" (2012, 4)—persists as a dominant paradigm.[1] My hope is that this article may help to delegitimize its premises and thereby contribute to a new understanding of the ontology of living creatures which recognizes their ethical status and mandates responsive ethical treatment.

1983: A Mobilization for Animals occurred with 15,000 people assembled at four major locations targeting the Primate Research Centers.

I will argue that classical science and the objectivist epistemology established under its influence fail to appreciate the subjective nonphysical dimension—minds and consciousness—that inheres in all living creatures (and perhaps in all physical matter). A recognition of their subjecthood not only engenders ethical knowledge; that ethical knowledge, it is here claimed, *inheres in* and *emerges from* the communicative encounter between subjects. There is, in short, an ethical substrate—a cosmic sympathy, expressed through what I am calling *emotional qualia*, that comes alive or becomes apparent in communicative encounters between living entities. Just as qualitative properties, such as the taste, smell, or feel of a physical entity arise or emerge when a subject encounters that object, so emotional qualia emerge in one's encounter with another subject. The knowledge of that subject thus necessarily includes an ethical dimension.

Accounting for its inadequacy as an epistemology for living creatures, the epistemology entailed in classical science remains deficient in at least two other important, underlying respects: one, it fails to explain the behavior of subatomic particles or quantum reality; and two, it provides no explanation for the phenomenon of "emergence"—that is, how mind "emerges" from matter; how *qualia* or qualitative features emerge from *quanta* or physical objects; how, in short, *res cogitans* emerges from or relates to *res extensa*, to reprise familiar Cartesian distinctions—the basis for scientific epistemology.

An investigation of these two aporias, which I pursue in these pages, helps to lay the ontological groundwork for an enriched understanding of how sympathy—the root praxis in ecofeminist care theory—operates. Such refinement enables us to respond more adequately to critics of care and sympathy theory who question the basis upon which it rests, namely, the assumption that animals are subjects whose ethical voice is available and understandable, were humans to take the trouble to hear it—a position I have argued in a series of articles (Donovan 1990, 1996a, 2006, 2013).

While dialogical theory has been well developed as an ethical basis for human interaction by such figures as Martin Buber, Mikhail Bakhtin, Iris Murdoch, Simone Weil, and others, and extended to nonhuman creatures by Patrick Murphy (Murphy 1995; see Donovan 1996b); the line of thought developed by the numerous modern philosophers, whose views I treat in this article, extends the concept of dialogue—that is, conversation between two subjects—beyond human interaction to establish a "participatory epistemology" as the basis for human ethical knowledge of the natural physical world, including, of course, animals.

In what follows, I first examine the inadequacies of classical Cartesian/Newtonian science with respect to quantum physics, showing how there exists a substratum of reality—posited as mental and emotional—unexplained by classical objectifying theories. Next, I consider how they fail to explain the "emergence" problem. Both of these considerations lead to the conclusion that there is a communicative medium unseen and undetected by classical science, accessible by consciousness, through which qualitative information is transmitted, and for which a different alternative, "participatory" epistemology is required. Ontological reconceptions along these lines, philosopher Thomas Nagel has recently asserted, "will require a major conceptual revolution at least as radical as [Einstein's] relativity theory" (2012, 42; see also Orr 2013, 27). It behooves ecofeminist care theorists in animal ethics to remain abreast of these philosophical transformations, for a reconceptualized non-Cartesian epistemology is necessary, I contend, in order to substantiate claims about animals' subjecthood and to recognize the ethical communication their very presence, as living subjects, expresses in our encounters with them; namely, that they do not want to be slaughtered, eaten, tortured, exploited, or otherwise harmfully interfered with.

Consciousness, quantum ontology, and the neglected background

Ironically, the classical objectifying scientific view has been destabilized and disrupted from within science itself—in the field of quantum physics. Indeed, given the understanding we now have of the workings of physical reality on the most primordial, quantum level, it appears that the only knowledge now available to us—that is now possible—is not objective but *participatory*. The perplexing difficulties in determining the exact nature and behavior of subatomic particles has led many physicists to conclude that subatomic reality cannot be known apart from the human observer. As Werner Heisenberg, one of the pioneer quantum physicists, explains in *Physics and Philosophy*, definitive identity of subatomic phenomena occurs only through the "act of observation": "The transition from the 'possible' to the 'actual' takes place during the act of observation ... as soon as the interaction of the object with the measuring device, and thereby with the rest of the world, come into play" (1958, 54–5).

That the observing self and the measuring instruments are necessarily implicated in the material being observed or measured is the so-called "Copenhagen Interpretation" of quantum physics proposed by Danish physicist Niels Bohr. In his "Discussion with Einstein" (1949) Bohr explained: "we are just faced with the impossibility, in the analysis of quantum effects, of drawing any sharp separation between an independent behavior of atomic

1983. Feminists for Animal Rights begins creating a slide show to illustrate visually the negative connections culture makes between women and animals, entitled "The Re-Presentation of Women and Animals."

objects and their interaction with the measuring instruments which serve to define the conditions under which phenomena occur" (quoted in Barad 2007, 308). As Karen Barad explains in *Meeting the Universe Half-Way: Quantum Physics and the Entanglement of Matter and Meaning* (2007), there is an "ontological entanglement [between] objects and agencies of observation" (309). "Our knowledge-making practices are material enactments that contribute to, and are part of, the phenomena we describe" (249). In short, "we are not outside observers of the world ... [W]e are part of that nature we seek to understand" (184). Quantum physics therefore requires a "participatory epistemology" (de Quincey 2002, 18) because "the observer cannot ... be ... separated from the object being investigated ... [such that] the so-called observer is actually a *participator* ... [and] any particular probability becomes an actuality *only when observed*" (29).[2] Quantum theory implies, therefore, a "profound participatory relationship between knowing and being" (de Quincey 2002, 151), because an observer's consciousness *participates* in the realization of a particular given entity (*realization* in two senses: the becoming real of the entity [ontological] and the understanding of and knowledge of the entity by the observer [epistemological]).

Another baffling feature of quantum physics that invites, indeed requires, a reformulation of the traditional epistemology of classical science is the phenomenon of "*quantum non-locality*," whereby a measurement on "one member of a quantum pair of particles" "simultaneously 'influence[s]' the other member" which is at a distance from "the original one." In other words, "some kind of 'influence' ... connects ... across spacelike-separated events" (Penrose 1994, 245). As this distance can be millions of miles, it is indeed unfathomable by ordinary standards, suggesting a cosmic communicative interconnectedness hitherto inconceivable—and incomprehensible in terms of classical Newtonian physics. As one theorist notes, "the phenomenon of *quantum nonlocality* suggests a strong (by now unexplainable) entanglement between all [subatomic particles] ... [T]wo quantum objects Q_1 and Q_2 can be distant for light-years and nevertheless, when Q_1 collapses then Q_2 collapses at the same moment ... Since superluminal velocity [faster than the speed of light] violates all known laws of physics, there has to be another (experiential?) connection between Q_1 and Q_2," (Spät 2009, 168). In other words, there has to be some sort of communicative medium to which both particles attach.

Sir Arthur Eddington, a distinguished astrophysicist, theorizes that there is an "unknown background" within which and from which this communication occurs. Humans' "physical knowledge," he writes, "is based on measures" which register measurable phenomena, that is, material phenomena that have features—objective data—that can be detected by our instruments and observation. "The physical atom," in other words, "is ... a schedule of pointer readings." But, he specifies further, "the schedule is ... attached to an *unknown background*" (Eddington 1928, 259, emphasis added). What we detect through our instruments and observations are "measure-groups resting on a shadowy background that lies outside the scope of physics" (Eddington 1928, 152).

Simone Weil in her essays on quantum physics came to a similar conclusion. In "Classical Science and After" [La Science et nous] (1941), Weil invokes the concept of *attention* (later developed in her ethical theorizing about "attentive love"—see Weil 1977; Donovan 1996a) to claim that physics fails to pay attention to the "negligible," which is "nothing other than what has to be neglected in order to construct physics" (Weil 1968, 31), "for physics is essentially the application of mathematics to nature at the price of an infinite error" (34). In another piece, "Fragment: Foundation of a New Science" [Du fondement d'une science nouvelle, n.d.], Weil returns to "the notion, fundamental in physics, of the *negligible*" (Weil 1968, 81). "[S]omething is missing in these images [detected by scientific instruments]—and that something ... is the presence of the whole surrounding universe," for the images "are not existing things but abstractions" (81).

Eddington, however, maintains that the "shadowy" "unknown background" behind the images or "pointers" is nevertheless accessible through the mental phenomenon of consciousness. "[N]o one can deny that mind is the first and most direct thing in our experience, and all else is remote inference" (281). "[C]onsciousness" is "the only avenue to what I have called *intimate* knowledge of the reality behind the symbols of science" (340). Mathematics and physics are "extracted out of the broader reality ... It is in this background that our own mental consciousness lies" (282); indeed "consciousness has its roots in this background" (330). Eddington contends provocatively therefore that the unknown background is mental, "mind-stuff" (276), "something of spiritual nature of which a prominent characteristic is *thought*" (259). "[L]et us accept the only hint we have received as to the significance of the background–namely that it has a nature capable of manifesting itself as mental activity" (260).

More recently, in *Mind and Cosmos* (2012) Thomas Nagel echoed that the great inadequacy in classical science is the failure to explain or include mind.

> The great advances in the physical and biological sciences were made possible by excluding the mind from the physical world. This has permitted a quantitative understanding of that world, expressed in timeless, mathematically formulated laws. But at some point it will be necessary to make a new start on a more comprehensive understanding that includes the mind. (2012, 8)

1983. Following in the tradition of the Greenham Common Women's Peace Camp (started in 1981), the Porton Down Women for Peace and Animal Liberation occupies the area surrounding the Porton Down military testing site in England. In their public statements and their leaflets they protested the military industrial complex, vivisection, and provided a feminist analysis of the violence of the state.

Nagel thus foresees monumental reconceptualizations on the horizon, which will apprehend that "mind is not just an afterthought or an accident or an add-on, but a basic aspect of nature" (2012, 16).

The emergence problem

That qualitative expression is the primary means through which communication occurs in this mental medium is suggested when we reflect upon the emergence problem, the other phenomenon about which classical objectifying science fails to offer an explanation; namely, the question of how and/or if mind "emerged" or "emerges" from matter—which has plagued Cartesian theory from the onset. Likewise, how qualia "emerge" or relate to *quanta*, or objective physical entities, remains unexplained in the Cartesian/Newtonian I–it epistemology of classical physics. "[P]hysics describes the … spatio-temporal outline of things, but says nothing about the qualitative stuff" (Hartshorne 1937, 178).

Patrick Spät gives as an example of qualitative emergence the quality of saltiness and its relation to the physical molecule of sodium chloride. The molecule by itself is not salty. Saltiness only "emerges" or occurs when experienced by a tasting subject. "For there to be something having the property of saltiness one needs something that experiences this property" (Spät 2009, 162). Spät emphasizes that this does not imply subjectivism or Berkeleyan idealism. The saltiness "really *is* in the sodium chloride as an unrealized disposition—and with the intervention of an experiencing subject this *disposition becomes realized*" (163). The parallel to the quantum physicist whose observation occasions the quantum phenomenon to take or be expressed in a definitive state would seem to be apparent (though Spät doesn't make this connection). In both cases it is the presence of an experiencing subject, a conscious mind, that effects the determined reality. The conscious mind together with the observed object enable the emergence of the qualitative experience. Through this participatory dialogue a new intersubjective reality is created or emerges through the interaction between the two entities.

The panpsychist explanation of this otherwise inexplicable communicative ontology is that proposed by Eddington; namely, that there is a mental substrate to all reality, that "mind, or some mind-like quality, is present in all parts of the natural world, even in matter itself" (Skrbina 2009, 2). Both ends of the communicative link—observer/experiencer and observed/experienced (though subject-object terms are inappropriate; see below)—reside in a mental matrix through which the communication occurs. And it is through the communicative link itself, the moment of connection, that a new reality—saltiness in the above example—is co-created.

Freya Mathews, a contemporary philosopher who espouses a panpsychist position aligned with the process philosophy of Alfred North Whitehead and Charles Hartshorne, proposes a "subject-subject continuum" instead of "subject-object dualism" (Mathews 2003, 170). "Subjectivity," she argues, "is fundamental to the nature of reality" (7). As opposed to objectifying scientific knowledge, what is needed, she contends, is "a dialogical and participatory relation to the world" (6). Like many other theorists in this vein, Mathews derives an ethical mandate from this participatory episte-mology. "It is through encountering the world, making contact with its subjectival dimensions that we ... acquire a sense of spiritual kinship, which ... provide[s] the basis for a respectful and sympathetic attitude" (79).

Emotional qualia and the ethical response

Even more revolutionary in their implications for a reformulation of the relationship between epistemology and ethics, however, are the proposals put forth by Charles Hartshorne and more recently Patrick Spät, who maintain that an ethical response *inheres in* the "intimate knowledge" (to reprise Eddington's term) humans have of other realities. As saltiness emerges in the encounter between a conscious subject and sodium chloride molecules, so ethical awareness *inheres in, emerges from,* a subject's experience of a dog's whine. The physical basis for the dog's pain exists in and of itself,[3] corre-lating to the sodium chloride molecules, but its emotional expression—the whine—emerges as ethically actionable in the encounter with a registering subject of consciousness through the medium of sympathy.

Hartshorne, for example, stipulated, "In the whine of a dog, a listener intuits a feeling tone of displeasure *in the whine itself*" (Dombrowski 2004, 88). It is not a matter of interpreting "simulacra of feelings" (78) or of adding on "value" "after the fact of sensation" (88); it is rather a matter of experi-encing the feeling *with* the other subject, whether it be pain or joy—a reprisal of phenomenologist Max Scheler's concept of sympathy or *Mitgefühl* (see Donovan 1996 and further discussion below). Indeed, Hartshorne defines "sympathy" as "feeling *of* feeling" (Hartshorne 1980, 82). And in this sympathy—this "feeling *of* feeling"—resides ethical awareness. It is a matter of connecting to the substrate of "cosmic sympathy" (Hartshorne 1937, 316) that is expressed through (emotional) qualia.

Patrick Spät likewise draws ethical conclusions from this sympathetic participatory epistemology, arguing that unlike in classical Cartesian epistemology where the "is" is separated from the interpretive "ought," here the "ought" is *inherent in* the "is." That is because in experiencing another's pain we know inherently that the other wishes that pain to be salved because *we feel it* ourselves and *feel about it* this way (wishing it to end).

> We condemn harming [a] dog not just because we invent a moral principle and thus impose our projections onto physical reality; namely his experienced pain. It is his physical reality that exhibits an experiential reality which *shows* us that the dog wants to avoid the awkward pain. The bifurcation between "is" and "ought" is nothing but an artificial armchair principle. (Spät 2009, 170)

Panpsychism, Spät claims, "has the power to bridge the old chasm between the Scylla and Charybdis of 'is' and 'ought'" (170); rather, "following panpsychism" the two are not to be viewed separately. That is, "experiential reality is not just 'in my head' which observes and interprets a mindless and dead physical reality, but rather my experiential reality 'meets' another experiential reality ... while looking at a suffering human being [or] dog" (171). "Our experiential reality tells us how physical reality is like from the inside" (171), thereby evincing ethical knowledge or values: "If a living organism 'is' suffering we 'ought' to protect or help it" (171). The ethic is inherent in the experiential knowledge. A dog's whine, a cow's bellow, a horse's shying, a spider's flight, a snake's flinch, a bird's screech: these are emotional qualia whose ethical message is unmistakable. We feel them in our bones.

Communicating with animals requires this kind of "participatory epistemology," a kind of sympathetic caring alertness or attentiveness to the signs that are being communicated and an emotional "feeling-with" or *Mitgefühl* that is in itself a form of ethical understanding. While many maintain that the operation is one of analogy—that we understand others' mental and emotional states by analogy to our own, more illuminating in my view is the concept of immediate emotional understanding as here described by Hartshorne and Spät. Max Scheler similarly insisted, "we can understand the experience of animals ... [as] for instance when a dog expresses its joy by barking and wagging its tail," seeing this communication as reflecting "a *universal grammar* valid for all languages of expression" (Scheler 1970, 11). Although Scheler undoubtedly meant them metaphorically, the terms *grammar* and *language* unfortunately tend to connote a verbal, symbolic expression that falsifies the actual reality of the experience, which, as Hartshorne and Spät emphasize, is one of feeling the feeling of another subject at the same time through the emerged *emotional qualia*. It is not therefore a matter of analogizing the dog's expressive behavior to our own. We humans in fact do not express ourselves in an analogous way. We don't make verbal noises analogous to barking or jump around wagging our rumps when we feel joyful. Nor are we likely to feel joyful as often as dogs or to their degree of intensity. Nevertheless, we can understand their feelings; we *feel* their feelings. "The mortal terror of a bird, its spritely or dispirited moods, are intelligible to us and awaken our fellow-feeling [*Mitgefühl*]" (Scheler 1970, 48).[4]

Consciousness and living matter

While numerous metaphysicians have proposed a panpsychist view of material reality, science itself has in recent years come forth with evidence that tends to corroborate this perspective. We have seen how quantum physics invites a panpsychist interpretation—that there is some sort of "mind-stuff" inherent in physical reality. Equally suggestive, in my view, are recent developments in studies of living material—from cells to plants to nonhuman animals. Noted biologist Lynn Margulis has, for example, proposed in a recent article (2001) that cells in living bodies are indeed conscious. Her argument relies on her theory of "symbiogenesis"—that early life-forms were constituted when microorganisms were "ingested" but not "digested" (Margulis 2001, 59) in the eukaryotic cells that were the ancestors/antecedents of all living cells today. As certain of today's microorganisms are homologous to those "ingested" eons ago, and as these contemporary microorganisms are clearly "conscious" in the sense that they respond to and communicate with their environment, it stands to reason, she argues, that today's eukaryotic cells manifest a form of consciousness.

An earlier study done at the University of Wisconsin, "'Decision'-Making in Bacteria" (Adler and Tso 1974) similarly imputes a form of conscious awareness in microorganisms. When *E.coli* bacteria—one-celled organisms—were presented with two options, they "chose" the one more favorable to their survival. While this reaction could be dismissed as a mechanical chemical response, the fact that the bacterium was presented with both a "repellent" and an "attractant" in equal quantities at the same time meant that some sort of "decision" had to be made and that that determination had to be transmitted to the flagella which then reacted accordingly (i.e. the bacterium moved). Even if the process is a chemical reaction, how is that explainable without reference to some sort of mental stratum

1984. Feminists for Animal Rights publish their first newsletter.
1984. A conference on "Women, Nature, and Science" is held in Sweden.

or medium?—another instance of the "emergence" problem. How does matter (chemical molecule) otherwise convey information (nature of the repellent/attractant materials) to other matter (bacterium molecules) which is conveyed to other matter (flagella molecules), and then converted to energy (movement of flagella)? One has to posit some sort of mental communicative medium in which qualitative valuing occurs.

A correlative philosophical tradition accords mental or spiritual reality to plants, on the assumption of some sort of subjecthood or consciousness therein, beginning in the nineteenth century primarily with German scientist Gustav Fechner whose *Nanna, oder Über das Seelenleben der Pflanzen* [Nanna, or On the Soul-life of Plants] appeared in 1848 (parts of which have appeared in English translation as *Religion of a Scientist* [1946]).[5] Fechner argues that the "inward and vital relation between the forces and activities of the [plant's] individual cells" (Fechner 1946, 206) constitutes the plant's "soul." "Everything one can reasonably require as essential to the expression of the soul is to be found in the plant as well as in the animal ... If one will not venture to deny that the plant has life, why deny it a soul?" (168) ["was man füglich als wesentlich zum Ausdruck, der Beseelung fordern könnte, bei Pflanze sich noch eben wohl als bei Tiere vorzufinden ... ; wagt man doch nicht einmal, der Pflanze Leben abzusprechen, warum spricht man ihr doch Seele ab" (Fechner 1908, 7–8)].

While such views will be dismissed by many as anthropomorphic or an example of the so-called "pathetic fallacy" (Hartshorne criticizes the latter concept, pointing out that equally and perhaps more distortive is the "apathetic fallacy" [Dombrowski 2004, 93]), recent scientific investigations suggest that there may be something to Fechner's intuitions. In 2011 a team of scientists at Ben-Gurion University in Israel published a study that indicated a pea plant, when "subjected to drought conditions communicated its stress to other such plants with which it shared its soil, ... prompting them to react as though they, too, were in a similar predicament" (Marder 2012, 9). The information was communicated biochemically through the roots (how that information "emerged" from the plant's molecules is, of course, unexplained by classical science). Thus alerted, the plants were able to "activate appropriate defenses and adaptive responses" when themselves faced with drought conditions, thus indicating that they had stored the information communicated from the afflicted plant in a sort of memory (9). From this one may conclude that though "they do not have a central nervous system, [plants] are capable of basic learning and communication" (9). A plant is, in short, "not only a *what* but also a *who*—an agent in its milieu, with its own intrinsic value or version of the good" (9). These examples appear to reinforce the thesis that some sort of sympathetic communicative continuum exists in which cells, microorganisms, plants, animals, and humans operate.[6]

Participatory science

Philosopher Henryk Skolimowski urges in *The Participatory Mind* that while

we cannot be wolves … we cannot be mountains … through our acts of empathy which stem from our deep participation in all that is, we can … so identify with the well-being of other beings that this empathy becomes an act of reverence … [which leads to] tolerance, … protection, … preservation, … care and … love. (Skolimowski 1994, 377)

Seeing American biologist Barbara McClintock's work as exemplary, Skolimowski calls for a new, ethical science: "we have to make a transition from objective consciousness to *compassionate consciousness*," thus laying the groundwork for a "new participatory science" (1994, 165).

McClintock, a Nobel-prize-winning geneticist, is known for her intense involvement with the plants she studied, an involvement which yielded an extraordinary knowledge of the intricate workings of corn-plant genetics, leading her to see the plants she worked with as subjects, as "ensouled" (Keller 1983, 204). "Over the years," Evelyn Fox Keller notes, "a special kind of sympathetic understanding grew in McClintock, heightening her powers of discernment, until finally, the objects of her study have become subjects in their own right" (Keller 1983, 200). An "organism" for her became seen as "a living form … object-as-subject" (200).

In explaining her method to Keller, McClintock noted that while examining the minutest of living particles, chromosomes,

I found that the more I worked with them the bigger and bigger [they] got, and when I was really working with them I wasn't outside, I was down there. I was part of the system. I was right down there with them … It surprised me because I actually felt as if I were right down there and these were my friends … As you look at these things, they become part of you. (Keller 1983, 117).

One begins to feel a "real affection" for them (Keller 1983, 117).

In emphasizing the "subjecthood" of the plants, McClintock stressed that for the scientist it is a matter of being receptive to what the living tissue is telling one—not forcing it to behave in prescribed ways. Rather one must be patient and attentive: one must "let it come to you"; one must "hear what

1984. The Bloodroot Collective's *The Second Seasonal Political Palate: A Feminist-Vegetarian Cookbook* published. In it they explain: "Our relationship to the earth and her creatures is the same relationship we must have with each other as sisters: when we hurt the earth we hurt each other; when we create with the earth we create with each other."

the material has to say to you" (Keller 1983, 198). "You let the material tell you where to go" (125). Moreover, one has to be attentive to each individual organism and get to know it individually, as plants even in the same species differ widely and have their own unique characteristics.

> No two plants are exactly alike. They're all different, and as a consequence, you have to know that difference … I start with the seedling, and I don't want to leave it. I don't feel I really know the story if I don't watch the plant all the way along. So I know every plant in the field. I know them intimately, and I find it a great pleasure to know them. (Keller 1983, 198)

This kind of "intimate knowledge," Keller notes, yields a "feeling for the organism" (198) that is intellectual, emotional, visceral—a shared participatory knowledge stemming from the fact that both observer and observed are living beings who operate within the same communicative medium and can therefore connect and exchange information on that basis. For McClintock, Keller explains, it was a matter of perceiving and animating the "soul" in her plants. "The ultimate descriptive task [she felt] … is to 'ensoul' what one sees, to attribute to it the life one shares with it" (204).

Failing to appreciate the subjecthood of living creatures—that is, their nonphysical being, which may include mental or spiritual life, consciousness, experiential capabilities, and/or communicative sensitivity—relegating this subjectivity to the status of "negligible," to reprise Simone Weil's term, is to overlook the phenomenon of ethical emergence described by Hartshorne and Spät and to remove the rationale for ethical obligation embedded therein—the knowledge that when "a living organism 'is' suffering we 'ought' to protect or help it" (Spät 2009, 171). We are emotionally and ethically engaged with that creature by dint of belonging to the same communicative mental medium. As we *feel* their pain, ethical awareness emerges—in a sense it becomes *our* pain—and through that connected feeling we are thereby moved to ethical action.

Transgenic engineering and humanist hubris

In concluding, I would like to consider the practical implications of this participatory ethic, considering its application, for example, in the newly evolved field of transgenic engineering. I do so by contrasting the view offered here to that proposed in "material feminism," a new vein in feminist theory which developed partly in reaction against the "discourse" turn in postmodernist thought that construed the natural world—indeed all reality—as ideologically constructed. Because it attempts, inter alia, to take into account recent developments in quantum theory—in particular, the notion that we humans are inherently "entangled" in the physical, natural world—material feminism would seem to be complementary to the view developed here. However, some material feminists, in particular Karen Barad and Donna Haraway, have drawn different ethical conclusions from the fact of human "entanglement"—conclusions that are antithetical to the participatory ethic proposed here. Their view is that because we are already entangled in physical reality and since physical reality is always in process, we are licensed to further tinker with, or mess with, what we're already a part of and what is already changing anyway. On this argument one may justify human rearrangement of nature by any means, including, for example, genetic engineering.

In *Meeting the Universe Half-Way* (2007) Barad implies that we humans have this kind of license because of our inherent entanglement. Riffing on T. S. Eliot's question (in "The Love Song of J. Alfred Prufrock") "Do I dare disturb the universe?" (394), Barad says, not only is the answer "yes" but we have an ethical responsibility to do so. Since we are "part of the world in its becoming" we have an obligation to "the possibilities of becoming." As the "world and its possibilities for becoming are remade with each moment" (390) we are entitled to participate in that remaking. We should heed the "ethical call" "to take responsibility for the role that we play in the world's differential becoming" (396). On this view Barad defends biomimesis and transgenic engineering as an example of "reconfiguring entanglements" (384) and therefore contributing—presumably positively—to the "world's becoming."

Barad developed this legitimation in response to a forceful criticism of transgenic engineering by environmentalist Janine Banyus as "a biological transgression of the worst kind" (367) (in a piece aptly entitled "The Height of Hubris"). Barad refutes this critique by joining Donna Haraway in dismissing "environmental activists" who object to such bioengineering as guilty of "reifying a notion of nature based on purity" (369).

Haraway in *Modest_Witness@Second_Millennium: Female Man©–Meets–OncoMouse™* (1997) makes the, in my view, outrageous—indeed slanderous—charge that critics of transgenic engineering are engaging in the type of purity fetishism seen in such ideologies as Nazism and racism. "Transgenic border-crossing signifies serious challenges to the 'sanctity of life' for many members of Western cultures, which historically have been obsessed with racial purity" (Haraway 1997, 60). (Never mind the egregious guilt-by-association assumed here: being a member of a Western culture means that one is obsessed with "racial purity.") Haraway hammers home her point by insisting that in the criticism of transgenic engineering one "cannot help but hear ... a fear of the alien and suspicion of the mixed ... a mystification of kind and purity akin to the doctrines of white racial hegemony" (61).

The concept of purity is, of course, anathema to all reflective persons in a post-Holocaust world for reasons I presume to be obvious. However, aside from being a shocking slur on serious ethical critics, it is a red herring in this discussion. No environmentalists that I'm aware of—least of all ecofeminists—espouse anything that remotely suggests purity. On the contrary, they have repeatedly emphasized diversity with an emphasis (especially in the ethic-of-care tradition) on paying attention to the particular in all its uniqueness—in other words, the practice advocated by Barbara McClintock and theorized by Marti Kheel (2008, 2–7, 11–12, 17) and others. And certainly the theorists I have considered in this article are not only *not* reifying nature (which entails objectification or turning into a thing, the Latin root of the term being *res* or "thing"), as Barad accuses, they are *subjectifying* it. Ironically, if any practice reflects an obsession with purity, it is precisely that of transgenic engineering, which purports to "better" or purify nature according to human design—a practice to which Barad and Haraway offer qualified endorsement.

The position taken by Barad and Haraway in short fails to respect the living reality of the physical and natural world and its creatures—their subjecthood; fails to consider that they might have an ontological status equal to humans'; fails to consider that they might have something to say that merits consideration in the matter of human intervention in "their" world. The Barad/Haraway position fails in short to put the onus on humans to make an effort to know that world and those creatures on the level of *intimate* knowledge described in this article, and to react ethically in accordance with that knowledge.

Instead, what we have in their position is a rebirth—despite posthumanist pretensions—of humanist hubris: a reiteration of the presumption that we humans have the right to remake the world in ways we preconceive.[7] The justification has changed; it is no longer because we are human that we have this right but because we are "entangled"; but the net effect remains the same: humans operating on, dissecting, destroying, and rearranging an objectified natural world in accordance with their wishes and ideas. It is in short, ironically, merely an extension of classical scientific praxis, rooted in Cartesian epistemology.

As Freya Mathews explains, a panpsychist ethic necessarily involves a critique of that praxis, seeing it as "a transgression and intrusion, a violation" (2003, 76). Instead, a participatory ethic involves "encountering the world, making contact with its subjectival dimensions ... [thereby] acquir[ing] this sense of spiritual kinship, which provides the basis for a respectful and sympathetic attitude"—one that is "nonintrusive" and doesn't "dismantle" the world and "rebuild it according to ... abstract designs" (79)—which is, of course, exactly what transgenic engineering does.

A participatory epistemology of the type described in this article entails, in short, as ecofeminist theologian Rosemary Radford Ruether asserted years ago, not "the conquest of an alien object" but a "conversation of two subjects," a realization that "the 'other' has a 'nature' of her own that needs to be respected and with [whom] one must enter into conversation" (Ruether 1974, 195–6). Her side of the conversation—her feelings—must be taken into account, must be *part* of any human decision-making about the natural world and her creatures. By paying attention to information communicated through emotional qualia we realize the subjectivity of other beings—other "subjects of feelings"—and in that awareness lies caring, respectful, compassionate ethical knowledge.

Notes

1   In 2012, prompted no doubt by developments in cognitive ethology (see Ristau 2013), 25 leading neuroscientists issued "The Cambridge Declaration on Consciousness in Non-Human Animals" (accessible at http:fcmconference.org), which concludes that animals "possess ... the neurological substrates that generate consciousness." The Cambridge Declaration is thus a promising sign that the Cartesian model may be weakening as the reigning dogma; however, the primary research in this field is oriented toward identifying "neural correlates of consciousness," thus remaining within a dualistic paradigm. As Thomas Nagel notes, "psychophysical reductionism," the idea that "the mental can be identified with some aspect of the physical, such as patterns of behavior or patterns of neural activity," remains a dominant and "entrench[ed] ... world view" (Nagel 2013). As intimated in this article, it is my view that more radical reconceptions are called for than availed in this world view. Thanks to Lori and Carol for alerting me to the Cambridge Declaration.

2   The phenomenon being referred to is the so-called "collapse of the wave function" (de Quincey 2002, 29), in which the act of measurement or observation causes the quantum wave-particle phenomenon to "collapse" into a definite state. "The act of measurement ... forces a system into a definite state and place at a

given time ... [such that] the initial mixed wave function 'collapses' into a precise state" (Lederman and Hill, 2011, 189, 225; see also Barad 2007, 280). Before the "collapse," the wave function—a mathematical equation—only indicates probabilities of time and location; it is only upon interaction with the measuring instrument and observer that a precise entity in time and space is determinable (or exists).

3   That the dog's pain would also be registered internally through the same sympathetic medium is a position taken by Hartshorne, inter alia. Hartshorne argues that there is a scale of being from simple units to the more complex, from, for example, "electrons in a cell, cells in a vertebrate." "The higher include the lower individuals *as such*—i.e. without reducing them to the role of mere 'matter'" (Hartshorne 1937, 123). Within the human body (indeed within the body of all multicellular organisms) the relationship between mind and body, he contends, is "essentially one of sympathy between the radically inferior sentient cells and our human consciousness" (29). In a chapter entitled "Mind and Body: Organic Sympathy" in his magnum opus *Beyond Humanism*, Hartshorne proposes that there is a "sympathetic interaction" (200) between body cells and mind: "our feeling is a sympathetic participation in cell-feelings" (198), and vice versa. We are "directly, intuitively sensitive to bodily cells which are also sensitive to us" (209). David Ray Griffin explains, "*We* feel pain ... because we perceive our bodily parts sympathetically" (Griffin 1998, 110).

4   Hartshorne similarly insists on the communicative expressiveness of birds, seen as subjects capable of independent creative action. In his book on birdsong Hartshorne argues against behaviorist views: animals, he says, are "aesthetically enjoying and not merely behaving creatures" (1973, 2). A complementary theoretical approach to animal communication is that developed by phenomenologists Edith Stein (1966), Ralph Acampora (2006), and Elizabeth A. Behnke (1999). See also Kenneth Shapiro (1989).

5   Numerous literary figures from Margaret Cavendish (1972, 67, 69) to William Wordsworth (1967; Hartshorne 1980, 81) and Sarah Orne Jewett (1881, 168–82) have espoused an animist view of plant life—a position seemingly ratified in recent Israeli experiments with pea plants.

6   This is not to argue that we have comparable ethical obligations to bacteria or plants as to animals; however, the same ethical bottom line obtains in all these cases: we should not harm or kill living creatures who are not harming or threatening our wellbeing or the lives of those for whom we are responsible (see Donovan and Adams 2007, 4).

7   For a similar, incisive critique of Haraway, see Weisberg 2009, 45–55.

Eros and the Mechanisms of Eco-Defense

pattrice jones

5

esire drives everything. Arising in our animal bodies, eros impels us to stretch and strive for what we want. What we want, most of all, is connection.

Rooted in patriarchal pastoralism, globalized via colonialism, serving the aims of capitalism, and furthered by slice-and-dice-style science, the hegemonic economy of (re)production and consumption is the catastrophic antithesis of exuberant eros. It persists by damming and diverting eros along with rivers.

Sparked by rioting street queens and enacted in explicit solidarity with the Black, Chicano, Native American, and women's liberation movements of that era, the fabulous gay liberation movement of the 1970s has devolved into a fairly conservative movement that asks for only reactionary "rights" like marriage and military service. We need a theory and praxis of animal liberation that resuscitates the queer spirit of rebellious and generous connectedness.

To be fully realized, the ecofeminist ethos of care (Kheel 1993) must be nourished and informed by eros. But "love don't come easy," as Diana Ross and the Supremes once sang. Eros can't be hurried, ordered around, or expected to march in anything like a straight line. To resuscitate eros, we must understand its queer ways.

1985. Marilyn French's *Beyond Power* published.
1985. Anti-fur activists break away from Greenpeace and found "Lynx." They engage in the first direct harassment of women wearing fur coats. (Lynx painted them red).

Steps to an ecology of eros

"The diversity of modes of singing amongst birds is so great that it defies explanation"
—*C. K. Catchpole and P. J. B. Slater, Birdsong (2008, 234)*

"We don't only sing, but we dance just as good as we want."
—*Archie Bell, introducing himself and the Drells on the 1968 recording of "Tighten Up"*

The leaves of Bruce Bagemihl's (1999) 750-page encyclopedic account of animal homosexuality teem with "wuzzling" dolphins, "necking" giraffes, and "cavorting" manatees, not to mention "aquatic spiraling," "sonic foreplay," and a form of sexual stimulation known as the "genital buzz"—and all of that just in the few pages devoted to an overview of the "dizzying array" of ways that nonhuman animals court and show affection to one another (13–18). In all, Bagemihl carefully reviews the documented accounts of same-sex courtship, affection, pair-bonding, parenting, *and sex*—did I mention "mounting, diddling, and bump-rumping"?—among the members of some 300 species of mammals and birds.

Zoo visits, televised nature programmes, and storybooks featuring stereo-typically gendered characters teach us to think about other-than-human animals as relentlessly heterosexual despite "the much more prevalent sex diversity among living matter" (Hird 2004, 86). It's not just pop culture that gets it wrong. Bagemihl (1999) also documents the long, sorry history of scientific obliviousness, bewilderment, and heterosexist hubris in the face of same-sex sexuality among other-than-human animals. From the ethologist who decried the moral degeneracy of butterflies to the wildlife biologist who evicted a same-sex couple from the nest they had built together so that he could give it to a heterosexual pair, the litany of wrongs and wrong-headed writings is leavened only by the unintentional humor of the sometimes surreal extremes to which scientists have gone in order to avoid seeing (much less naming) the queer eros right in front of them.

Before we ascribe the bemusement of those scientists entirely to the mutually reinforcing junction of ignorance and bigotry, consider this: The fungi known as *Schizophyllum Commune* swap genes by touching and have as many as 23,000 mating types (or, as we like to call them, sexes), thereby preventing "selfing" in a species in which any individual can both give and receive genetic material in order to produce progeny (Casselton 2002). Confused? That's my point. Not only does non-reproductive sexuality flower in a variety of forms, but sexual reproduction itself occurs by means of a "remarkably diverse" (Fraser and Heitman 2004) array of strategies.

But—wait!—there's more: not only some plants but also some animals reproduce by various asexual means, including parthenogenesis. As Catriolina Mortimer-Sandilands and Bruce Erickson (2010) note in the introduction to their important anthology, *Queer Ecologies*, this "diversity of asexual modes of reproduction as well as several multi-gendered ones … appear to defy dominant, dimorphic accounts of sexual reproduction altogether" (12).

Alaimo (2010) and others have commented on the inadequacy of our conceptual categories in the face of all of this. As biologist J. B. S. Haldane famously opined, "the universe is not only queerer than we suppose, it is queerer than we *can* suppose." So it's not surprising that eighteenth-, nineteenth-, or even twentieth-century scientists unwittingly assimilated their observations of same-sex behavior into dualistic schemas they themselves couldn't see (because they seemed like reality).

Of course, just because mushrooms swap genes by brushing against each other is no reason to presume that we could or should do the same. Just because marsupials are also mammals does not—alas—mean that we can bound around with infants in our pouches. Nonetheless, this survey of the variety of (always embodied) animal eros offers us much more than an antidote to the still-too-common misconception that homosexuality is unnatural. First: Things we can't imagine right now might still be possible. And: We too may be queerer than we suppose.

People have courted, demonstrated affection, constructed households, and raised their children in a blooming profusion of different ways. And, while there's some doubt that we deserve the sobriquet of *sapiens*, there's no doubt at all about the *homo*. We not only sing to our same-sex sweethearts in almost as many languages and styles of music as there are varieties of birdsong, we also dance together in configurations that don't fit within the boy-girl two-step of the Western square dance—and not just in urban discos.

Since errors and erasures in this realm are almost as endemic as those concerning same-sex sexuality in nonhuman animals, a brief survey is in order. While gay or lesbian *identity* constructed in contradistinction to the relatively recently invented category of "heterosexual" is fairly new and geographically bounded (Katz 2007), same-sex erotic *behavior* has been "virtually universal in human societies" (Drucker 1996, 75), sometimes in defiance of cultural norms but often with cultural toleration or approval. From traditional marriages between African women (Morgan and Wieringa 2006) to casual sex "play" between Pakistani men (Khan 2001)—neither considered particularly queer by the participants—expressions of same-sex desire continue to thrive even in regions where they are repressed or reputed not to exist.

None of this is news. Or, rather, none of this should be news, since the evidence—like that of animal homosexuality—has been hidden in plain

1985. For the first time the Annual Conference on Women and the Law includes a workshop on animal rights.

1985. Dian Fossey murdered in Rwanda, Zaire.

view. For example, in Africa, where rock-solid scientific certainty of the absence of indigenous homosexuality delayed appropriate AIDS-prevention interventions for many years and where several states still justify anti-gay legislation with the idea that homosexuality is alien to the continent, indigenous same-sex sexuality "is substantially documented in scores of scholarly books, articles, and dissertations in a wide range of academic disciplines, in unpublished archival documents … in art, literature, and film and in oral history from all over the continent" (Epprecht 2008, 7).

In the Americas, culturally condoned expressions of same-sex sexuality were so common among indigenous peoples that this was frequently cited by invading Europeans as justification for cultural genocide (Galeano 1992; Katz 1976; Smith 2005). In some North American native cultures, "two-spirit" people—those believed to have both male and female aspects due to their gender presentation and/or sexual orientation—were not only tolerated but esteemed (Roscoe 1988). Similarly, "there are striking examples of the recognition and acceptance of forms of same-sex desire in the history of important parts of Asia" (Sanders 2005, 32), including China, India, Japan, and Java. Local varieties of homoeroticism in Thailand and elsewhere in Asia-Pacific were and remain truly diverse, confounding not only heteronormativity but also easy explanations about what "homoeroticism" might mean (Jackson 2001; Wieringa, Blackwood and Bhaiya 2007).

As Chou (2001) cautions about China, homosexualities of the past ought not be simplistically romanticized. Some traditional patterns of same-sex sexual behavior, in that region and elsewhere, occurred within and were patterned by social inequalities that today we would condemn as unjust. Some historical reports of same-sex eroticism record its repression. Nevertheless, whether they have been esteemed, approved, tolerated, used, abused, or repressed—and however they have thought of or denoted themselves—people who sometimes or exclusively have sex with members of their own sex have existed in virtually every human population. We're just that kind of animal.

Many—perhaps most—of the people who have sex with partners of the same sex do not think of themselves as homosexual or even bisexual. At the same time, terms denoting homosexual identity (or something like it) abound. While the globalization of *gay pride* has led to the importation of the terms *gay* and *lesbian* and variants thereof into numerous languages (Katyal 2002), local terms both old and new express local ways of enacting and thinking about same-sex sexuality. Again, it is necessary to resist idealization. While some contemporary same-sex practices challenge social inequalities or are integrated into egalitarian communities, others conform to—and perhaps reinforce—binary notions of gender and patriarchal conceptions of power (Blackwood and Wierenga 2007).

Nonetheless we can revel in the linguistic and conceptual creativity deployed in the service of same-sex sexuality. In China, some activists have repurposed the term *tongzhi*—a Chinese translation of a Soviet-era term for comrade, constructed from *tong* (same) and *zhi* (spirit, goal, or orientation)—as a way of denoting themselves in a manner congruent with cultural values (Chou 2001). In Uganda, the previously derisive term *kuchu* is now claimed with pride (Tamale 2007). In the United States, many Native Americans have embraced the term *two-spirit* as "an expression of our sexual and gender identities as sovereign from those of white GLBT movements" (Driskill 2004).

Diversity in self-identification also flowers within populations. Thailand's national lesbian organization uses the term *ying-rak-ying* (women who love women) to describe its constituency; at the same time, many Thai women who do love women reject that term in favor of *tom* (short for tomboy) or *dee* (short for lady), because these better express their sense of themselves within the rigid gender system in which their sexualities operate (Blackwood and Wieringa 2007).

Which brings us to gender. Recently, Czech archaeologists unearthed neolithic skeletal remains of a biological male who had been buried—4,500 to 5,000 years ago—in the manner in which females of that time and place usually were interred (Karpova 2011). Considerable diversity in ideas about gender has been documented in the human cultures that have arisen in different times and places since then, both in terms of whether gender is mutable (Nanda 2000; Ramet 1996) and how many genders there might be (Davies 2006; Roscoe 2000). Communities also have varied in the ways that they have coped with or explained persons who don't quite fit into any of the categories created by their culture's gender system (Epprecht 2008).

The multiplicity of sexualities and exuberant diversity of gender expression among *Homo sapiens* suggests that some measure of flexibility—or, at least, a capacity for and tendency toward variability—is intrinsic to our species. The rigid enforcement of a two-gender system that goes hand in hand with rigid insistence on relentless heterosexuality is an artifact of a particular set of circumstances, seeming natural only because of its tendency to reproduce itself.

Televised nature programs tend to portray evolution as an urgent quest in which every animal attempts to spawn as many offspring as possible. Some scientists, too, have implicitly defined evolutionary success as reproduction of the individual rather than the survival of the family, population, or species. Some "evolutionary psychologists" attribute virtually every characteristic and behavior to the reproductive imperative implicit in this presumed law of nature.

1986. Susanne Kappeler's *The Pornography of Representation* implicitly links the status of women and the status of animals as objects whose role is to subjectify men.
1986. Alice Walker's "Am I Blue" published in *Ms.* Magazine.

But, if incessant reproduction is the law, we've got an awful lot of animal scofflaws. In some species, only a few individuals even attempt to reproduce. In many species, females—for whom reproduction is often a physically perilous affair—actively avoid pregnancy by means of a variety of strategies (Bagemihl 1999).

Natural selection in the sense of helpful genetic mutations passed along to offspring who then disproportionately survive to reproduce certainly does occur and certainly does explain many evolutionary changes. But exclusive focus on the reproductive success of individuals ignores the interlocking material and social circumstances that also evolve. It's not *quite* right to say that organisms adapt to their environments, since organisms are *part of* their ecosystems, which themselves constantly change as their participants evolve (Oyama 2000). Furthermore, natural selection acts upon not only physical traits but also behaviors, many of which are transmitted by social learning (Avital and Jablonka 2000). Moreover, cultural and physical traits often co-evolve.

Natural selection acts upon groups as well as individuals. The overall fitness (or lack thereof) of the social group—family, flock, tribe, or troop—significantly influences likelihood of individual survival. Some circumstances—such as high rates of predation—do mandate that everyone at least attempt to reproduce for the population to survive, but in most instances the problem faced by populations is the opposite. The long-term survival of any group requires that its population be calibrated to the availability of resources. Same-sex sexual activities (and non-reproductive heterosexual erotic activities) allow for pleasure, bonding, and other benefits without risk of reproduction. Hence, queer eros enhances group fitness.

Adults who don't reproduce help groups in other ways. Homosexual pairs adopt orphaned offspring in many species. Adult animals who do not reproduce—including but not limited to exclusively homosexual animals—tend to contribute more to their social group than they take out. In all social circumstances where adults cooperate to provide protection and resources to juveniles, those who do not reproduce contribute without withdrawing. Those adults also have more energy to devote to activities that benefit the group, whether these be writing operas or looking out for predators.

The simplistic view of natural selection and the European mindset in which it emerged both presume a struggle for scarce resources as the precondition of life. This way of thinking about the world makes sense in its own ecological context. Even after the decimation of the plagues and purges of the centuries just past—the traumatic impact of which must still have been reverberating in European psyches—the population of Europe remained too high to be satisfactorily supported by its deforested and depleted ecosystems.

Within such simultaneously barren and crowded surrounds, the Hobbesian view of each against all must have seemed self-evidently true.

But, in fact, plenitude rather than scarcity is the norm in nature. Most animals—like most people, prior to the successive waves of conquest and consequent population explosion that have globalized the environmental crises from which they arose—spend comparatively little time securing food and shelter, with plenty of time left to play. Virtually every human culture has produced music, visual art, and sport of some sort. All of these—along with non-reproductive sexuality—can be seen as exuberant uses of the abundant energy that shines down from the sun every day (Bagemihl 1999).

Bagemihl contrasts his theory of biological exuberance with the notion that homosexuality serves as a natural check on population, but I see those ideas as complementary and mutually reinforcing. The exuberant upsurge of queer eros keeps populations in check, thereby setting the stage for even more exuberance. Suppression of queer eros thus injures populations, and their enveloping ecosystems, as well as individuals.

Reproduction and its discontents

In zoos and vivisection labs, animals are assorted into boy-girl pairs and forced to mate if they do not do so willingly. Often, this involves breaking up same-sex pair bonds or preventing females from fleeing unwanted penetration. Likewise, in animal agriculture—whether on factory farms or family farms—everything depends on reproduction. From the electro-ejaculation of bulls to the confinement of fragile "broiler breeder" hens with heavyweight roosters made sex-mad by starvation, numerous cruel and unusual strategies ensure that no farmed animal opts out of compulsory heterosexuality.

Even animal lovers join in the superintendency of animal sexuality. Dog lovers who decry puppy mills still feel free to decide whether, when, and with whom the canines under their control will partner. Animal advocates pursue the laudable goal of reducing animal homelessness and execution by demanding that all companion animals be deprived of reproductive freedom rather than by abolishing the for-profit traffic in dogs and cats. Animal sanctuaries, with similarly pragmatic rationales, routinely prevent their residents from choosing to reproduce.

Meanwhile, homosexual or transgender behavior or identity remains a risky endeavor for people in many places. In some 35 countries, homosexual behavior remains a crime punishable by imprisonment. In South Africa—the first country in the world to enshrine non-discrimination on the basis of sexual

1987. First national gathering of the Green Movement includes the formation of an Animal Rights Caucus that begins to prepare an Animal Rights platform for the Green Movement.

orientation in its national constitution—lesbians still confront an everyday threat of "corrective rape." Here in the United States, Mercy for Animals cofounder Nathan Runkle was gay-bashed nearly to death only a few years ago.

The structural function of homophobia is the maintenance of the man-on-top binary gender system (Pharr 1988). That system dates back to the days when both daughters and dairy cows were the property of males who presumed the right to force females—whether they be called wives, slaves, or livestock—to bear more or different offspring than they would otherwise choose. Patriarchy and pastoralism both require fairly relentless preoccupation with and control of reproduction (and, hence, sexuality). The traditional pastoralism from which today's factory farms evolved *necessarily* involved hands-on control of the reproduction of captive animals (Patterson 2002). Tools and tactics first used to gain and maintain total control over nonhuman animals were adapted for use with human slaves (Spiegel 1996). The process of conquest by which men who viewed women, land, animals, and people of other races as property created a globalized economy also carried homophobia around the world.

In today's topsy-turvy world in which Uganda's Christian president condemns homosexuality as a foreign import despite the fact that Christianity comes from elsewhere while the indigenous Langi people allowed marriage between men and biologically male *mudoko dako* people (Tamale 2007), it may be important to repeat that queer eros flowered in a variety of flavors prior to the era of European colonialism and imperialism. While cultures varied in their attitudes towards homosexual behavior, toleration appears to have been the norm (Epprecht 2008), including in places where homophobia is now most marked (Galeano 1992).

What happened? First, European Christians brought their atypical antipathy to homosexual behavior with them when they invaded the Americas, in some instances using native sex and gender norms as excuses for cultural genocide (Galeano 1992; Smith 2005). Obversely, Europeans were vested in the notion that Africans were like animals—and, hence, relentlessly heterosexual. "The prevailing prejudice was that Africans were uncivilized and close to nature … The emerging consensus on homosexuality thus required that Africans conform to the expectation of a supposedly natural heterosexuality" (Epprecht 2008, 40). The mechanisms by which African sexual diversity was denied or suppressed eerily echo those by which homosexuality among nonhuman animals was elided from the record for so long. Anthropologists failed to see, refused to record, hesitated to publish, or explained away instances of same-sex sexuality of which they were aware.

These colonial cover-ups were in many instances compounded by post-colonial leaders eager to avoid any suggestion of the luridly "exotic" spectacles of African sexuality promulgated by National Geographic and the like (Epprecht 2008). Many of these leaders were men eager to enjoy the subordination of women facilitated by homophobia (Tamale 2007). Meanwhile, the post-colonial wave of trade globalization brought commodified conceptions of "gay," "lesbian," and "transgendered" identity to regions where indigenous cultures had other ways of conceptualizing same-sex activity (Katyal 2002). Rooted as they are in European ways of thinking about identity, these may or may not prove to be useful to queer people elsewhere—or to our shared struggle to bring ourselves into balance with the biosphere—but they certainly have provoked a fresh wave of homophobia-fueled violence in many places (Blackwood and Wieringa 2007).

How did "gay" or "lesbian" get to be nouns instead of adjectives (or, even better, verbs) anyway? In short, the same Enlightenment ideas that brought us scientific racism engendered a way of thinking that led eventually to the categorization of people on the basis of what we now call sexual orientation. "The rise of evolutionary thought in Charles Darwin's wake generally coincided with the rise of sexological thought in Richard von Krafft-Ebing's" (Mortimer-Sandilands and Erickson 2010, 7). As we've seen, sexual selection is but one aspect of natural selection. Yet, perhaps because of the preoccupation with reproduction implicit in patriarchy/pastoralism, that element of evolution became the center of evolutionary theory. In consequence, "sex became a matter of fitness, and individual attributes could now be evaluated based on their apparent adaptiveness to an organism's reproductive capacity" (Mortimer-Sandilands and Erickson 2010, 8).

That was bad news for eros. The medicalization of homosexuality arose at the same time, and from the same train of thought, as eugenics and scientific racism. Many of the medical assaults on people who were (or were perceived to be) either homosexual or non-gender-normative "occurred in the context of Race Hygiene and Race Betterment movements"; not only queers but also deaf, disabled, or dark-skinned people were considered "literally, biological enemies of the human species" (McWhorter 2010, 76). All of these efforts to improve the species presumed and sought to preserve the position of *Homo sapiens* at the top of an imagined hierarchy, the scientific rationalization of which continues to make speciesism feel logical even to those who now know that evolution is not an upwards affair.

Along with delusions of grandeur and a perverse preoccupation with reproduction, the process of conquest that has led us to the present juncture exported an ethic of exploitation and a fantasy of infinity. That ethic, and the fantasy that enables it, have since been codified into the economic rules by which European powers still force the rest of the world to play.

1988. *Rape of the Wild: Man's Violence Against Animals and the Earth* by Andrée Collard with Joyce Contrucci published.

1988. Karen's Davis's "Farm Animals and the Feminine Connection" published in *Animals' Agenda*.

Capitalism demands and indeed requires incessant growth—new markets for new goods, which must come from somewhere—in order not to collapse. Unlike economies in which participants cooperate to trade fairly, capitalism is mathematically unbalanced by the removal of profits into private pockets and thus requires constant infusions of fresh resources. Thus, it requires not only incessant reproduction—whether of factory farmed chickens, assembly line automobiles, or worker-consumers to build those cars and eat those birds—but also the diversion of desire. Every natural impulse, whether for self-expression or social contact or—yes—sex, must be detoured toward the purchase of some product (for which, of course, one must earn the money to buy). And so now queer eros, where it is not still actively suppressed, faces the same dispiriting fate as heterosexual romance—the destination wedding!

Having integrated virtually every place on earth into its economy of empty rapacity, late consumer capitalism has run out of places to go for fresh supplies of worker-consumers. Everything now depends on getting everybody to buy *more*—which is of course the opposite of what everything actually depends on.

Seven billion people now stand on an overheated planet. "Humans have already changed the biosphere substantially, so much so that some argue for recognizing the time in which we live as a new geologic epoch, the Anthropocene" (Barnosky et al. 2012, 57). Climate change comes down to "patterns of human behavior, particularly over-population and over-consumption" (Oscamp 2000), both of which follow directly from the suppression and diversion of eros in all of its exuberant diversity.

Preoccupation with reproduction characterizes most present-day human cultures. Reproduction remains an obligatory duty to family and community even as we confront the catastrophic environmental effects of overpopulation. Parenthood remains conflated with adulthood in many minds. This "repro-centric" (Mortimer-Sandilands and Erickson 2010, 11) logic is both cause and consequence of suppression of queer eros. If reproduction is the paramount goal, then non-reproductive eros must be suppressed; if non-reproductive eros is suppressed, eros will seek satisfaction in socially sanctioned reproduction and consumption.

Unless...

Consciousness of lost limbs

In 2001, the mass trial of 52 Egyptian men arrested for dancing together at a floating nightclub rightly drew international attention to the ongoing persecution of homosexuals in that country (Hawley 2001). But let's notice something else: these were men who *knew* they faced prosecution for homosexual activity. And they were *dancing*. To *disco*. On a *boat*.

From "Fiddler on the Roof" to "Mississippi Masala" (not to mention "Romeo and Juliet" and all of its remixes), pop culture thrums with tales of young lovers defying parental prohibitions to follow their hearts. In the real world, eros often leads lovers of all ages to disobey even more powerful authorities. Here in the United States, in South Africa, and elsewhere, laws prohibiting miscegenation aimed to maintain oppressive racist regimes, but men and women of different races persisted in partnering, the threat of jail notwithstanding. All around the world to this day, same-sex couples come together despite real and menacing threats ranging from social ostracism to the death penalty.

"Beneath the paving stones—the beach!" Like half-forgotten dreams, anarchist slogans like that Situationist gem from the 1968 student rebellion in Paris pop up on walls and burst from the mouths of black-clad teenagers smashing shop windows with baseball bats. Tattooed survivors of childhood sexual abuse tuck battered copies of the CrimethInc Ex-Workers Collective Manual *Days of War, Nights of Love* into backpacks, to read while sitting in endangered trees. Vegan punks bake cupcakes for each other, just to bring some sweetness to their struggles.

Eros is right there—ready—to show us where we need to go and give us the energy to get there. Yet eros is also so easily deadened or misdirected. Eros can help us save ourselves and each other, and quit wrecking the planet along the way, but to do so it must be deliberately cultivated.

It might seem counterintuitive to pursue the aim of checking human hubris by cultivating human eros. But true eros, unlike plastic pleasures purchased from profiteers, is both enlivening and relational. Eros is exuberant, sometimes jumping up when and where you least expect it. Eros begins in the body but always reaches outward, seeking connection.

By "eros" I mean not only physical love and sexual desire but also what Black lesbian feminist poet and activist Audre Lorde (2012) called the "sharing of joy, whether physical, emotional, psychic, or intellectual, [that] forms a bridge between the sharers" (56). Eduardo Galeano (1992) reports that among the Maya of Colombia, the ancestors of whom rose up against the sexual constraints imposed by the Conquistadors, the word for sex is "play." Eros *is* playful. Queer eros is both cause and consequence of a happy state of affairs in which life is *not* a grim struggle to reproduce in the face of scarcity but, rather, joyous usage of the surplus energy that shines down from the sun every day.

Because eros is inherently surplus and always oriented outward, genuine eros is always generous. We share smiles with sweethearts and give gifts and

1988. *Woman of Power: A Magazine of Feminism, Spirituality, and Politics* published a "Nature" issue. With Ingrid Newkirk on animal rights, Moran's guidelines for raising children as vegetarians, a feminist critique of the notion of animal "rights" that argues that the best way to help animals is by adopting broad ecofeminist values.

Figure 5.1 At VINE Sanctuary, cross-species care-giving is common. Here, lamb ALFie offers a tender greeting to arriving calf Maddox. They have since become best friends. Photograph © by Kathy Gorish for VINE Sanctuary.

kisses to beloved others. It *feels good* to do this. Eros awakens feelings of all kinds—including the ones that ought to be telling us this is an emergency and giving us the energy to do something about it. As Lorde wrote in her classic essay on the uses of the erotic, "as we begin to recognize our deepest feelings, we begin to give up, of necessity, being satisfied with suffering and self-negation, and with the numbness which so often seems like their only alternative in our society" (58). Our feelings are fuel, and the good news is that they are renewable sources of energy.

Eros arouses not only desire but also curiosity, creativity, and courage. It has abidingly proven to be more powerful than guns or governments in motivating human behavior. Inherently ecological, because it begins in our animal bodies, eros undammed and undiverted will flow in the direction of biospheric balance.

Greta Gaard (1997) points out that "dominant Western culture's devaluation of the erotic parallels its devaluations of women and of nature" and that "these devaluations are mutually reinforcing" (115). As with all of the intersections of oppression, the upside of seeing the junction is recognizing opportunities for intervention. And so, when we nurture wombats, women, or weeds genuinely and with generosity, we also are cultivating eros. But, if we hope to use eros to reanimate ourselves while animating our environmental efforts, we will have to come to better terms with our own animality.

The animal problem

The Eurocentric logic of mind-over-matter tells us that we should transcend our animal bodies by means of our very fine minds. That's the same logic, however, that divides and conquers the world into male-over-female, white-over-black, and straight-over-gay. My favorite flavor of ecofeminist theory extends the feminist understanding of intersectionality to include earth and animals, deepening our analyses of race, gender, and other social constructs along the way.

Neither homophobia nor speciesism (nor any other ism) is a disembodied idea. They are *practices* (and accompanying rationalizations) that arose at particular times and places for particular purposes. Perhaps the most important purpose, for both of those and for sexism and racism too, is control of reproduction. Thinking about that intersection forces us to face not only our own animality but also our complicity in the ongoing subordination of other species by our own. That raises the question of how to go about animal liberation.

Neither we nor the other animals we propose to liberate are abstract entities. Actual animal liberation is all about bodies—theirs *and* ours—and is therefore all about eros.

Without eros, ethos risks slippage into the realm of disembodied abstraction. Suppression of eros is suppression of our animal selves and is thus antithetical to the project of animal liberation. Suppression of eros severs us not only from our desires but also from others, deadening our feelings and relationships in the process. It's difficult to imagine how a liberatory ethos of care could be adequately enacted by beings who are cut off from themselves and others.

What would an *erotic* ethos of care bring to the project of animal liberation? First, eros is always embodied and therefore always actual. Animals don't care about our pretty ideas or pure intentions—what matters is what actually happens. An ethos of care rooted in eros would therefore mandate that care be actually enacted, that our ideas interact with that practice, and that both theory and praxis be constantly adjusted in response to what actually happens.

Next, eros is all about desire. Different animals want different things. Salmon want streams that haven't been dammed or diverted. Frogs want unpolluted ponds. Chickens want *out* of those battery cages. Dogs want other dogs. All of these desires are located in bodies. Their frustration is felt physically. So, again, this brings us back to the actual. But also, the animal rights movement as it is currently constituted does not, in my view, make an

1988. Marti Kheel presents a workshop on the connection between ecofeminism and animal liberation at the Greening of the West conference.

adequate effort to wrestle with that diversity of desires, preferring to focus on rights that are most important to animals like us (e.g. legal liberty). An ethos of care rooted in eros would mandate a much more thorough reckoning of animal desires and a consequent (and continuing) adjustment of aims and tactics.

Third, eros is all about relationships. An ethos of care rooted in eros would therefore mandate that such deliberations flow from, insofar as possible, real relationships with the animals in question.

Which brings me to my expanded conception of the organic intel-lectual—let's call it my theory of the veganic intellectual. As conceptualized by Gramsci (1971), the organic intellectual—a person who conceptualizes and articulates the ideas of a class of people in which she is enmeshed—is essentially a function of the social group, both growing out of and acting upon the group. Whether or not they have formal education, organic intel-lectuals both learn and teach in the context of active engagement with the struggles of their group, whether that group be an economic class or some other aggregation.

Roosters have helped me not only to think through several important subjects but also to apply the resulting ideas in ways that help other roosters. When the first avian resident of what would become VINE Sanctuary turned out to be a rooster rather than a hen, he refused to allow my stereotyped ideas about roosters to define him or our relationship. This forced me to think deeply about where I got those ideas, which in turn led me to investigate the role of animals—real and imagined—in the social construction of gender. Similarly, a bonded pair of male foie gras factory refugees provoked me to commence a series of workshops at which participants considered the inter-sections between queer and animal liberation.

Then, in flew a group of two dozen roosters who had been living together for years. They schooled both me and the young orphans from factory farms in the methods and morals of flock life. Their habit of sleeping in the trees rather than in buildings eventually led our sanctuary to be the first in which chickens rewilded themselves. That wouldn't have happened if we hadn't decided to listen to the roosters themselves—who expressed themselves very clearly using their voices and bodies—on the question of how to balance freedom against safety from predation.

Our sanctuary was the first to figure out how to rehabilitate former fighting cocks. I say "our sanctuary" deliberately because this was a collective effort. The process involves not only soothing but also socializing these abused birds. I certainly could not have conceptualized that process without having first been taught about roosters by roosters, and the process cannot be implemented unless there are roosters willing and able to model the social behaviors that former fighters must learn in order to resolve conflicts without injury.

The veganic intellectual, then, plays the same role as the organic intellectual, but for a group that includes nonhuman animals. The veganic intellectual does not claim to be "the voice of the voiceless," but rather recognizes and listens to animal voices. The veganic intellectual—I think of Karen Davis (1995) with chickens and Lori Gruen (2009) with chimps— thinks in conjunction with nonhuman animals, exercising both empathy and careful observation, and then shares any arising ideas with people who don't have the same opportunities for communion.

The return of the repressed

Desire drives everything. It's easy to maintain patriarchy once you've tricked little girls into dreaming of their wedding days, and it's not so hard to control the working class if a preponderance of grown men are addicted to pornography and flat-screen TVs. Conversely, it can be difficult to engender progressive change while wild eros is dead-ended into such socially constructed cravings.

According to a warning recently published in *Nature* by more than 20 researchers in biology and allied fields, we seem to be "approaching a state shift in Earth's biosphere" (Barnosky et al. 2012, 52) wherein "the biological resources we take for granted at present may be subject to rapid and unpredictable transformations" (57). Reductions in *both* "world population growth and per-capita resource use" (57) will be necessary if we hope to avoid or even mitigate the coming cataclysmic changes.

Scientists haven't had much luck in using rational argumentation to persuade people to change our patterns of resource consumption (much less our mania for reproduction). Maybe we'll get lucky (in both senses of that phrase) if we focus on feeling instead, cultivating queer eros in all of its manifestations, including not only love among animals but also topophilia and biophilia. That project will depend on our ability to put people in touch with their most heartfelt desires (which won't tend to be wedding dresses or artisanal cheese), and that in turn will require us to embrace our own animality, including its queer eros.

1989. Barbara Noske's *Humans and Other Animals: Beyond the Boundaries of Anthropology* published.
1989. Donna Haraway's *Primate Visions: Gender, Race, and Nature in the World of Modern Science* published.

Acknowledgments

While I wrote these words, many of the ideas arose as a function of processes of collective cognition. I both acknowledge and thank all of the participants in all of the "Queering Animal Liberation" workshops and discussions I facilitated between 2002 and 2010 as well as all of the students in my 2011 and 2012 "Cultural Politics of GLBT Sexuality" classes at Metropolitan State University. Thanks are also due to Carol Adams, Rebecca Barry, Greta Gaard, Lori Gruen, and Miriam Jones for useful comments on drafts of this chapter or portions thereof. Words cannot quite convey the ways that I am also indebted to the nonhuman animals in whose presence I have tried to imagine what real liberation might look like and how we might get from here to there together. This chapter goes out to all of the former egg factory inmates who have collectively taught me to remain ready to be surprised by what turns out to be possible.

Figure 6.1 *Self-Portrait With Chicken* by Sunaura Taylor, 6″ x 4″, 2012, pen, paper and color pencil.

In this drawing an atypically bodied woman is shown in profile sitting naked with legs outstretched. Her back parallels the vertical right side of the paper while she faces towards the left, hairy legs stretched out in front of her. The figure is looking upwards. Perched on her leg is a plucked chicken who is also in profile. The chicken looks in the same direction as the human woman.

Both of the figures are wrong. This particular chicken cannot do the thing we most want birds to do, the thing that inspires us, that stirs our envy—she cannot fly. Plucked and awkward, she is lacking feathers, too long in the legs, and her wings have been reduced to nubs. We wonder whether the bird escaped in the midst of being processed for meat or for feathers, and we wonder what wounds she has been left with.

The woman is wrong too. She is unclothed, but not posing for our titillation. Her skinny, hairy legs align her with nature, and against culture. Like the chicken who doesn't fly, this woman's legs seem useless for walking. Her hands taper off into small nubs, perhaps limited in what they can grip and manipulate, but concave and cupped and ready to receive.

They are wrong as individual specimens and doubly wrong as a pairing, but they are not alone and not ashamed. The chicken is a familiar, a consort to the woman, just as the woman is a familiar, a consort to the chicken. They are denuded, open, waiting for something from above. The nature of that something is unexplained and undepicted, but hangs over everything, a cloud cover.

Maybe they should be afraid, but they are not.

This is an image of contemplation and companionship; of bodily analogies; of vulnerabilities; of questions about whether similarities outweigh differences.

Interdependent Animals: A Feminist Disability Ethic-of-Care

Sunaura Taylor

6

Being cared for

Feminists have long recognized the importance of inter-dependence and caring. Whether critiquing the ways in which caring for "dependants" has historically been placed upon the backs of women (and especially women of color), or drawing attention to an ethic-of-care (the ways in which caring should play a vital role in conceptions of justice), feminists have a long tradition of theorizing a worldview that understands humans (and often nonhumans) as being inter-dependent beings who care and rely on each other. However, where feminist theory has done a lot to theorize what it means *to care,* there has been less said about what it means to be *cared for.*

I have had a complex relationship to care. As a disabled person I both espouse a philosophy of interdependence (of which care is a vital component), while simultaneously resisting the narrative that care (especially in the form of goodwill, charity or the kindness of people's hearts) will somehow allow me to live a more liberated life. Being "cared for" can be stifling, if not infantilizing and oppressive (as of course can being in the role of the carer). In her article "Building Bridges with Accessible

1989. Marjorie Spiegel's *Dreaded Comparison: Human and Animal Slavery* published.

Care: Disability Studies, Feminist Care Scholarship, and Beyond," Christine Kelly writes: "theoretical work in disability studies implicitly and explicitly positions care as a layered form of oppression that includes abuse, coercion, a history of physical and metaphorical institutionalization, and a denial of agency" (Kelly 2013, 786). Historically, disability rights advocates have declared that we do not want to be cared for, we want rights, justice and an accessible society that does not limit or make impossible our involvement and contributions.

However, over the years there has been an emergence of feminist disability studies scholars and others who have tried to bridge these complications and create a theory of care that recognizes the value, but also the oppressive histories, of being both cared for and a carer (Kittay 2002; Nussbaum 2006; Kelly 2013). An important contribution this work can offer is a consideration of what it is that those who historically have been viewed as being cared for, those who have been labeled as dependants or burdens, contribute to their relationships, society and the larger world.

Theories of care and interdependence have also manifested themselves in a number of ways within conversations around animal ethics, particularly a feminist ethic-of-care toward animals. This work understands animals and humans as entangled in interdependent relationships, recognizing that animals are often vulnerable and dependent upon human care. In *The Feminist Care Tradition in Animal Ethics*, Carol Adams and Josephine Donovan write that animal rights theory "presumes a society of equal autonomous agents," however, as they argue, "Animals are not equal to humans; domestic animals, in particular, are for the most part dependent on humans for survival—a situation requiring an ethic that recognizes this inequality" (Donovan and Adams, 2007, 6).

Within a feminist ethic-of-care framework, the dependency and vulnerability of animals (particularly domesticated ones) adds to our responsibility toward them. Where many animal advocates have viewed animals simply as vulnerable victims who need protecting (often declaring themselves to be a "voice for the voiceless"), a feminist ethic-of-care offers a framework of justice that has the potential to complicate conceptions of dependency (perhaps in a similar vain to disability studies), to understand animals not as dependent beings with no agency, but rather as vital participants and contributors to the world.

Echoing certain core philosophies in disability studies about who is given voice and subjecthood, Adams and Donovan write: "Some ethic-of-care theorists emphasize that our attention be directed as well to what the animals are telling us—rather than to what other humans are telling us about them" (Donovan and Adams, 2007, 4). Although various questions exist as to how to listen to animals, an ethic-of-care nonetheless radically moves toward asking how we can help and care for animals without paternalism and infantilization. In a similar vein, philosopher Lori Gruen's work on entangled empathy challenges us to consider how our empathetic responses to others (specifically nonhuman animals) can help us not simply to sympathize with their suffering, but to consider what an individual animal wants, needs and is communicating.

Gruen writes: "Being in ethical relation involves, in part, being able to understand and respond to another's needs, interests, desires, vulnerabilities, hopes, perspectives, etc. not simply by positing, from one's own point of view, what they might or should be but by working to try to grasp them from the perspective of the other" (Gruen 2013, 224). The sort of empathetic understanding and paying attention that Gruen and Adams and Donovan call for seems deeply relevant to conversations around disability as well, particularly to conversations around intellectual disability. I want to be clear here that I am not comparing animals and disabled humans; instead I am recognizing that to understand another being who does not communicate in ways able-bodied/able-minded humans have historically valued, we must pay attention to individuals—learning from them so that we can recognize their agency and preferences. This is a crucial step in moving conversations about animal and disability liberation away from limited narratives about suffering and dependence to more radical discussions about creating accessible, nondiscriminatory space in society in which individuals and their communities can thrive.

Disability studies can add to the work being done by feminist animal ethics scholars by offering a critique of negative conceptions of dependency, and an understanding of interdependence that moves beyond mutual advantage, challenging us to consider what it is that animals are saying, wanting and contributing—while trying to consider what it is like for them to be cared for.

Disabled, domesticated, and dependent

It is generally accepted that disabled people are dependent. We are dependent on carers for our physical wellbeing, and often dependent on the government for our economic wellbeing. It is also generally accepted that domesticated animals are dependent: they rely on human beings for feeding, shelter, health care, often even with birthing and aid with intercourse. Wild animals rely on us as well, albeit in a very different way—they are vulnerable to human decisions that involve their habitats, their food sources, whether they as

1989. Two books with stories, reflections, and theory about women's relationships with animals published: *With a Fly's Eye, Whale's Wit and Woman's Heart* and *And a Deer's Ear, Eagle's Song and Bear's Grace*, edited by Theresa Corrigan and Stephanie Hoppe.

individuals can be hunted or poached, and sometimes even whether their species will survive into the future.

My libertarian grandmother once told me I should be grateful for everything I get as a disabled person, because I'd "die in the woods" if left to my own devices. What she meant is that if put in a "natural state" there would be no question of my complete and utter dependence—I would quickly starve unless someone kindly decided to share their berries with me, or (as my grandmother would have it) gave me some meat.

These are fighting words for a grandma (she was quite a character), but her basic thesis is actually widely accepted. The notion that disabled people only survive out of the goodness of other people's hearts has a long and widespread history. The point that my grandmother missed though, is that my able-bodied siblings would also eventually die in the woods if left all alone with no human support or tools. They might make it for longer than me, but odds are they'd go pretty quick.

Domesticated animals are also confronted with this line of thought. They are understood as man-made, unnatural, and as utterly dependent and unfit for the wild. Various environmentalists, animal welfarists, and animal advocates have presented domesticated animals this way—as tragically, even grotesquely, dependent. Disabled people and domesticated animals are burdened with many people's stereotypes about what it is to be unnatural and abnormal, as well as assumptions about the indignity of dependency.

 However, the truth is, all of us are dependent. We human beings begin life dependent on others and most of us will end life dependent on others. Yet dependence often becomes an excuse for exploitation and has extremely negative connotations—no one wants to be dependent.

In American rhetoric there is a strong emphasis on independence and self-sufficiency. America is the country where everyone has the opportunity to become "independent." Independence is perhaps prized beyond all else in this country and for disabled people this means that our lives are automatically seen as tragically dependent (Taylor, 2004). But how true is this?

Disability advocates argue that we are all dependent on each other. The difference between the way many disabled activists and scholars understand dependence and how the rest of society views it is that there is not so much emphasis on individual physical autonomy. In many ways independence is more about individuals being in control of their own services (be it education, plumbing, electrical, medical, dietary, or personal care), than it is about individuals being completely self-sufficient (Oliver, 1990); this is true not only for the disabled population, but for the population in general. Very few (if any) of us in the world are actually independent.

The negative consequences of dependency are largely made, whether through economic disenfranchisement, social marginalization, imprisonment, or societal, cultural and architectural barriers. In many ways the treatment of disabled people is merely a more pronounced form of the condition of other populations, as able-bodied individuals become dependent through socio-economic frameworks as well. The point is not that able-bodied people and disabled people are equally dependent, but rather that the dichotomy between independence and dependence is a false one.[1] Perhaps the distinction seems minor, but to many disabled people—a population persistently labeled as dependent and burdensome—a reminder that independence is far less clear-cut than is often presumed becomes vital.

However, it is also true that not all disabled people are able to be in control of or make decisions about their lives. As Michael Bérubé writes, "autonomy and self-representation remain an alluring ideal even (or especially) for people with disabilities" (Bérubé 2010, 102). Bérubé points to the fact that there are individuals who rely on others for all aspects of their survival and who lack not only physical independence, but the ability to make choices about their lives.

Dependency is real—but the point is that we all exist along its spectrum. The challenge is to understand dependency not simply as negative and certainly not as unnatural, but rather as an integral part of being alive. Eva Feder Kittay writes:

> I want us to see disability as sometimes (though not always) resulting in a dependency that is but one variety of a dependency that we have all experienced at some point and to which we are all vulnerable. Similarly, the care of the disabled person who needs to be assisted is but a form of care that many persons give to dependents of all sorts. My reason for eliding the differences in favor of the commonalities is that I believe that we as a society have to end our fear and loathing of dependency. We need to see our dependency and our vulnerability to dependency as species-typical. (Kittay 2002, 248)

As human beings our dependence on each other is actually a miniscule amount of our overall dependence. We are massively dependent on other animals and of course on our environments in ways that are impossible for us to really even fathom. Other animals are dependent on their communities, habitats, and ecosystems. None of us are actually independent. The whole planet is interdependent.

Despite this, disabled people become symbols of dependence. Where many able-bodied people can live in a delusion of independence (until of course they fall ill, hurt themselves, grow old, and so forth), disabled

le are often stigmatized as dependent and burdensome. Because of
our contributions to our families, communities, and cultures are often
negated.

Dependency has been used to justify the marginalization and exploitation
of both human beings and nonhuman animals. In *Animals Make Us Human*
Temple Grandin writes:

> I vividly remember the day after I had installed the first center-track
> conveyor restrainer in a plant in Nebraska, when I stood on an overhead
> catwalk, overlooking vast herds of cattle in the stockyard below me. All
> these animals were going to their death in a system that I had designed. I
> started to cry and then a flash of insight came into my mind. None of the
> cattle that were at this slaughter plant would have been born if people had
> not bred and raised them. They would never have lived at all. (Grandin and
> Johnson 2009, 297)

Dependency, it turns out, is a common argument for killing animals. The
animals we consume are dependent on us for their very existence. By eating
them we are doing them a favor.

Slowfood USA's "US Ark of Taste" program lists "over 200 delicious
foods in danger of extinction," the vast majority of which are animals—
heritage breeds.[2] As Josh Viertel of Slowfood USA told NPR, "You've got
to eat them to save them!" Their tagline reads "Saving Cherished Foods,
One Product at a Time."[3] In many ways the "eat them to save them" logic
of Slowfood USA is the pinnacle of consumer activism. By eating heritage
breeds, by literally consuming individual beings who are understood as
products, one is said to not only help small farmers, support local agriculture
and promote biodiversity, one can also save the animals themselves! But who
exactly are we saving?

Grandin and Slowfood USA use the extinction argument for different
means, Grandin to justify animal slaughter in general (including from the
largest producers), and Slowfood USA to support small farmers. But in both
cases this paradigm presents certain animals (namely domesticated ones) as
being dependent on their very own exploitation in order to live.

Dependency is a reasoning that has been used to justify slavery, patriarchy, colonization, and disability oppression. The language of dependency
is a brilliant rhetorical tool, as it is a way for those who use it to sound
concerned, compassionate, and caring while continuing to exploit those who
they are supposedly concerned about. Consider for example that farmer and
author Hugh Fearnley-Whittingstall argues that we must kill animals because
they are domesticated and thus will be dependent on us during their lives.
He writes:

> Of all the creatures whose lives we affect, none are more deeply dependent
> on us—for their success as a species and for their individual health and
> well-being—than animals we raise to kill for meat ... This dependency
> would not be suspended if we all became vegetarians. If we ceased to kill
> the domesticated meat species for food, then these animals would not
> revert to the wild ... The nature of our relationship would change but the
> relationship would not end. We would remain their custodians, with full
> moral responsibility for their welfare. (Fearnley-Whittingstall 2007, 16)

In a logic that is rooted in ableist notions of dependency as a negative and as burden, Fearnley-Whittingstall suggests that since we will still have a responsibility to these animals if we don't slaughter them (because they are dependent on us), we should eat them, as their dependent lives will be less worthwhile than their wild and independent counterparts.

Domesticated animals have in fact not only been presented as burdens who need to earn their keep, but in fact as "unnatural," environmentally damaging beings created by man. They are also repeatedly presented as dimwitted in comparison to their "natural" and "wild" counterparts. Ecofeminist Marti Kheel points out in her book *Nature Ethics* that environmentalist John Muir "expressed a common disdain toward domesticated animals when he described the dignity of wild goats as 'bold, elegant and glowing with life,' in contrast to domesticated goats who are 'only half alive'" (Kheel 2008, 5). This statement has a stunning parallel to the common sentiment that disabled people are incomplete, or as Jerry Lewis famously described, to be disabled is to be "half a person."[4]

In a similar vein, philosopher and environmentalist J. Baird Callicott wrote that domesticated animals "have been bred to docility, tractability, stupidity, and dependency. It is literally meaningless to suggest that they be liberated. It is, to speak in hyperbole, a logical impossibility" (Callicott 1989, 30).

The dependency of domesticated animals has often been presented in tandem with their supposed "stupidity," as if the fact that they cannot take care of themselves "in the wild" proves their dimwittedness. Of course numerous studies by animal behaviorists have shown that even when measured with human yardsticks, domesticated animals are hardly "stupid." Their intelligence is actually all the more striking when one considers the brutality the majority of domesticated farmed animals have undergone—being kept in environments completely devoid of mental stimulation for generations. However, the idea that it would be "meaningless" to support a population's liberation from exploitation, even if they were "dependent" and "stupid," is chilling.

Much of the hostility toward domesticated animals seems to come from the idea that they are unnatural; they have no natural state to be liberated to. Domesticated animals are often said to be destructive to the environment and at odds with natural habitats. I am not denying the horrendous impacts of animal agriculture on the environment and specifically global warming, rather an example of what I'm pointing to would be how many headlines blame these issues "on the cows," versus on humans' profound exploitation of them. Consider a more specific example, again from Callicott: "Domestic animals are creations of man. They are living artifacts, but artifacts nevertheless, and they constitute yet another mode of extension of the works of man into the ecosystem. From the perspective of the land ethic a herd of cattle, sheep, or pigs is as much or more a ruinous blight on the landscape as a fleet of four-wheel drive off-road vehicles" (Callicott 1989, 30).

One wonders what Callicott would say of dependent disabled people going for a stroll or hike in our power wheelchairs? Alison Kafer has written about ableism within the environmental movement and nature writings, and has shown how narratives of nature are persistently presented as being open only to those who can have an unmediated experience of "the natural." Kafer writes: "a very particular kind of embodied experience [is presented as] a prerequisite to environmental engagement ... to know the desert requires walking through the desert, and to do so unmediated by technology. In such a construction, there is no way for the mobility-impaired body to engage in environmental practice; all modalities other than walking upright become insufficient, even suspect. Walking is both what makes us human and what makes us at one with nature" (Kafer 2013, 132). Domesticated animals have also been seen as suspect—understood as unnatural "creations of man" who do not have natural interactions with nature, but instead are themselves equated with technology that harms nature.

However, as we saw with Grandin, Slowfood USA, and Hugh Fearnley-Whittingstall, a more popular—in some ways contradictory—argument sees animal dependency not as unnatural or bad for the environment (in fact domesticated animals are sometimes said to be essential for sustainable farming), but rather as justification for the continued use of animals for food. For example, Stephan Budiansky, author of *The Covenant of the Wild: Why Animals Chose Domestication*, writes: "one may argue that domesticated animals are degenerates that through dependency and excess kindness from humans have become weak and ever more dependent on the crutch of human care. But calling them 'degenerates' does not somehow mean they are less worthy of our consideration. If anything, their degeneracy ... argues for an even greater responsibility on our part" (Budiansky 1999, 123). Putting aside the almost laughable description of human "excess kindness" toward animals, and the prejudiced assumptions that disability, weakness, and dependency are inherently negative experiences, Budiansky gives us a picture of animal "degeneracy" that requires human "responsibility" versus contempt. However, this responsibility toward animals is expressed by raising, slaughtering, and eating them.

Considering all this, it comes as no surprise that dependency becomes an even more blatant issue when the animals being discussed are actually disabled themselves. In instances of animal disability, speciesism and ableism work together to render dependency an even more justifiable and righteous reason for exploitation. When an animal becomes disabled their continued contributions to their environments and communities are understood as less important, unessential or nonexistent, and their only contribution is said to be in their flesh. For instance, a recent controversy broke out at Green Mountain College over the slaughter of the school's two working oxen, Lou and Bill, who have tilled the school's land for nearly a decade. Green Mountain, which is known for its environmental and sustainable mission, voted to slaughter the two oxen for food as they were becoming disabled, aging, and unable to work. The decision was made after Lou stepped into a woodchuck hole, aggravating an injury in his leg. Lou already had "medical" issues that were making him unable to work, which along with the fact that the two oxen were aging led the school to decide to put them to use another way. As the assistant manager of the farm told the *New York Times*, "His quality of life is rapidly deteriorating, and this is the logical time to use him for another purpose." The focus on the two oxen's work value and productivity is telling—when animals are no longer earning their keep through working, their bodies must be put to work and made useful in another way. The *New York Times* piece quotes the farm's director: "'It makes sense to consume the resources we have on campus,' said Mr. Ackerman-Leist, who pointed out that the farm's purpose is to produce food in a humane and sustainable way, not to shelter animals. 'We have to think about the farm system as a whole.'" [5]

The ways in which romantic and conservative notions of self-sufficiency, productivity, and independence are entangled in contemporary discussions of animal welfare and sustainability is troubling. The work that Lou and Bill were able to perform as able-bodied animals justified (only for a time) their very right to live. As they became old and disabled the farm was adamant that they were not an "animal sanctuary." The people at Green Mountain College believed that Lou and Bill had only one way of earning their keep as disabled animals—as edible flesh.

The idea that some dependent individuals are less valuable and more justifiably exploitable because they are understood as burdens who offer nothing of value back to their communities, of course, has a long and troubling history for disabled humans as well. The sort of thinking that led to the crisis of Lou and Bill reveals limited ways of understanding what it means to contribute. Western concepts of mutual support and aid have largely been framed by philosophical traditions such as the social contract, which has privileged concepts of mutual advantage over other, sometimes less clear-cut, forms of support.

In her book *Frontiers of Justice: Disability, Nationality, Species Membership*, philosopher Martha Nussbaum shows how the tradition of the social contract has failed to provide substantial groundwork for justice for disabled people and nonhuman animals, as well as other vulnerable populations across the globe. The philosophical tradition of the social contract is a theoretical idea that emerged during the enlightenment, which tries to answer some of the questions of why individual, free, and rational people would choose to come together to govern themselves with laws in a society. The social contract argues that people who were roughly equal in strength and cognitive capacity chose to leave a "state of nature" and govern themselves for mutual advantage (Nussbaum 2006, 3). However, as Nussbaum writes, this influential theory "fails to address [disability, species membership, and nationality] as it assumes that in a 'state of nature' the parties to this contract really are roughly equal in mental and physical power" (Nussbaum 2009, 118). Of course, as Nussbaum points out, this assumption does not take into consideration physical and intellectual asymmetry between the disabled and able-bodied, and humans and nonhumans, as well as inequality between those who are born into wealthy nations and those who are not.

Nussbaum similarly shows how the social contract tradition's reliance on the idea of mutual advantage falls short when addressing disability and "species membership" as disabled individuals and animals don't necessarily offer mutual advantage *per se* and, in fact, in some cases may offer a disadvantage. Thus Nussbaum argues that a more complete theory of justice must challenge this tradition and include other more complex reasons for cooperation besides advantage, such as love, compassion, and respect.

An interesting parallel to Nussbaum's critique of the social contract is available in another contract theory: a co-evolution theory that authors such as Michael Pollan, Stephen Budiansky and Hugh Fearnley-Whittingstall use to justify meat eating. This theory says that human beings and domesticated animals have entered into a contract with each other that, like the social contract theory, is largely based on the idea of mutual advantage. The theory says that we have entered into a co-evolutionary pact with these species that gives us the responsibility to care for them in exchange for their services and flesh. To be vegetarian or vegan would mean abandoning those animals who are most dependent on us. Leaving them to their own devices, they say, would be a fate far worse than the dinner table (Taylor, 2011).

The theory says that if we look at things in evolutionary terms, domesticated animals are doing remarkably well. Their populations are high and spread all around the globe, and they have another species—humans—providing them food and shelter. The theory argues that the relationship of domestication, and the killing that goes along with it, is just as beneficial for the animals as it is for the humans. Once again, the argument is that if we didn't eat them, they wouldn't exist. As Pollan writes, "From the animals' point of view the bargain with humanity turned out to be a tremendous success, at least until our own time. Cows, pigs, dogs, cats, and chickens have thrived, while their wild ancestors have languished" (Pollan 2009, 120). Thus we have entered into a sort of social contract with these species, based on supposed mutual advantage; we provide and care for them and in return they feed our soil and give us their flesh. To stop eating animals would be to turn our back on this relationship and send these dependent domesticated creatures out into the wild only to die of starvation or be brutally killed by other animals.

As I have argued elsewhere (Taylor 2011), there are numerous problems and contradictions in this theory (for example, the high population that supposedly heralds these species' success is a consequence of intensive industrial animal farming, a practice these writers are all in opposition to). For our purposes, however, it is Nussbaum's critique of the power asymmetries in a state of nature that is useful here, as to argue that animals were on a level playing field with human beings when this supposed contract was being formed ignores the obvious fact that humans and animals have extremely varied mental and physical capacities. This bargain was not made between beings "roughly equal in mental and physical power," but between powerful human beings and more vulnerable animals. This contract was clearly written by the more powerful humans for their own interests: under this contract, humans benefit not only as a species but also as individuals, whereas animals "benefit" (if that word can be used at all) only as species, not as individuals.

When we argue that animals are dependent on their own slaughter for their very survival, we need to remember that it is we human beings who are choosing each and every time to slaughter them. When we say that by killing them we are letting them live we are declaring that these animals have only one purpose: to be used by us. We see them only as commodities that will be discontinued if there is no market for them.

In many ways the thinking behind the co-evolution theory is constructed around the idea of interdependence: domesticated animals and human beings have evolved together—animals help humans and we in turn help the animals, or so the argument goes. However, this interpretation of interdependence is lacking in a disability studies perspective. The interdependence that is discussed within this framework may be one of mutual advantage and support in some ways, but it also simultaneously devalues and takes advantage of those who are deemed weaker and dependent. What disability studies and a feminist ethic-of-care brings to the conversation is a more nuanced understanding of how to define mutual advantage and a much-needed analysis of what it means to be accountable to beings who are in many ways the most vulnerable. A disability studies perspective of interdependence is about recognizing that we are all vulnerable beings, who during our lives go in and out of dependency, who will be giving and receiving care (and more often than not, doing both), and that contribution cannot be understood as a simple calculation of mutual advantage. When we view animals through this lens we see that they contribute in countless calculable and incalculable ways, from fertilizing the soil to offering care and friendship. Lou and Bill, for example, clearly did contribute something very powerful to their community, which can be seen by the various people who offered them sanctuary and tried to save their lives.

Disability asks us to question our assumptions about who counts as "a productive member of society" and what sort of activities are seen as productive. It asks us to question the things we take for granted—our rationality, the way we move, the way we perceive the world. Animal ethics also requires a critical engagement with our assumptions about who is valuable and who is exploitable and a reimagining of what it means to contribute to the world.

Disabled, domesticated, and valuable

Surprisingly it is not only environmentalists and animal welfarists who argue that domesticated animals are unnatural, undignified, and dependent: animal advocates often argue this as well. Some animal advocates suggest that domestication has created beings who are so vulnerable to human exploitation that the only ethical solution is to stop breeding them until there are none. This includes farmed animals as well as cats and dogs. Domesticated animals are vulnerable as they are not only dependent on us, but have also been physically altered in unnatural ways that can be harmful to them—in other words, they are disabled.

Domestication has led to unquantifiable violence towards animals. As Sue Donaldson and Will Kymlicka write in their 2011 book *Zoopolis: A Political*

Theory of Animal Rights, "for many animal advocates, [domestication] is irredeemably unjust; a world in which humans continue to maintain domesticated animals cannot be a just world" (Donaldson and Kymlicka 2011, 73). Many animal advocates believe that the best thing for animals is to have nothing to do with us, but because domesticated animals are dependent on us for their survival, and cannot be separated from human society, they are better off not existing at all (Donaldson and Kymlicka 2011, 78). The reasoning behind an abolitionist argument for extinction is on one level very simple: if we stop bringing animals into existence, then they won't exist for human beings to exploit and make suffer. Some animal activists see the suffering and exploitation of domesticated animals as enough of a justification for their extinction. In a way I understand why the prospect of these species (that we have bred to be extremely defenseless against us) becoming extinct seems like the most responsible conclusion to the question of domesticated animals—after all we have done, why should we be trusted as caregivers?[6]

Like Donaldson and Kymlicka I argue that the extinction argument is troubling, especially when one considers just how much of it is based upon assumptions about dependency, naturalness, and quality of life (Donaldson and Kymlicka 2011; Taylor 2011). For example, consider this quote by animal advocate Gary Francione:

> Domesticated animals are dependent on us for everything that is important in their lives: when and whether they eat or drink, when and where they sleep or relieve themselves, whether they get any affection or exercise, etc. Although one could say the same thing about human children, the overwhelming number of human children mature to become autonomous, independent beings.
>
> Domestic animals are neither a real nor full part of our world or of the nonhuman world. They exist forever in a netherworld of vulnerability, dependent on us for everything and at risk of harm from an environment that they do not really understand. We have bred them to be compliant and servile, or to have characteristics that are actually harmful to them but are pleasing to us. We may make them happy in one sense, but the relationship can never be "natural" or "normal." They do not belong stuck in our world irrespective of how well we treat them. (Francione, 2012)

Francione's argument is strikingly similar to Hugh Fearnley-Whittingstall's statement presented previously, but they are arguing for completely opposite ends. It seems clear that the dependency and vulnerability of domesticated animals makes people on all sides of the animal debate profoundly uneasy. The ableist assumption that it is inherently bad, even unnatural, and

1991. Cathleen McGuire and Colleen McGuire convene a reading, discussion, and activist group called "EVE" (Ecofeminist Visions Emerging) which met for three years in New York City.

1991. PETA launches its "I'd rather go naked than wear fur" campaign; feminist critiques appear immediately.

abnormal to be a vulnerable dependent human being is here played out across the species divide—pointing to just how much ableism informs our ideas of animal life.

Wild animals in these narratives are romanticized—presented as the independent, natural subjects that Western philosophers have so long idealized. Domesticated animals on the other hand are seen as pitiable. In a parallel to the "better-off-dead" narrative of disability that views disability as worse than death, domesticated animals are viewed as "better-off-extinct." However, if we consider the often-gross misjudgments on quality-of-life issues that people make about disability, it becomes clear why it's so important to question assumptions about which lives are worth living. In fact it is impossible for me to consider the extinction view without conjuring up a history and legacy of eugenics.

Animal breeding and eugenics have an entangled and troubling history. Early eugenicists were inspired by the way animal breeds could be manipulated to have "better" traits. In the early part of the twentieth century Charles B. Davenport, a leader in the American eugenics movement and member of the American Breeders Association, a group devoted to furthering knowledge about genetics, heredity and breeding, described eugenics as "the science of the improvement of the human race by better breeding." Author Charles Patterson writes that Davenport "stressed the importance of people's genetic history and looked forward to the time when a woman would no more accept a man 'without knowing his biologico-genealogical history' than a stockbreeder would take a sire for his colts or calves who was without pedigree" (Patterson 2002, 83).

Eugenics had as its goal the idea of perfecting the genetic makeup of a population by ridding the genetic pool of "undesirable" traits, which were invariably linked to disability, race, and class. What we have done to farmed animals over the past century has already been a form of eugenics—genetically perfecting them for our purposes. We have selectively bred these animals over centuries to make them into better products, better specimens. Where in human beings perfection meant getting rid of "unwanted" characteristics such as disabilities, in animals it has often actually meant enhancing certain characteristics to the point that they easily could be classified as disabilities and deformities.[7]

Now that these animals are here with us, do we really want to enact another coercive force over their individual lives and species by leading them to extinction based upon assumptions that their lives are less worth living than wild animal lives? I find the idea that the solution to the wrongs of domestication is to erase the very populations whom we have harmed, extremely unsettling. Instead I want to ask, how can we dismantle the exploitative systems that have created these injustices in the first place? Part of this involves critiquing the idea that one's life is less valuable, worthy, or even enjoyable, if one is vulnerable and dependent.

I am not suggesting that animal advocates are promoting eugenics, but I am suggesting that all of us who identify as animal advocates think deeply about what it means to purposefully want beings to become extinct. We need to take seriously the possibility that just because an animal is not "wild" and may in fact be dependent on human care, that she also may simultaneously value and enjoy her life and may in fact still have agency. We need to consider that perhaps it is better that she exist than not exist. Although I am in complete agreement that what we have done to these animals is beyond egregious, as a disability scholar I am extremely weary of claims that domesticated animals are better-off-extinct because they are vulnerable and dependent.

Donaldson and Kymlicka write that "Dependency doesn't intrinsically involve a loss of dignity, but the way in which we respond to dependency certainly does." They offer the insightful example of a dog pawing at his bowl for dinner. "If we despise dependency as a kind of weakness, then when a dog paws his dinner bowl … we will see ingratiation or servility. However, if we don't view dependency as intrinsically undignified, we will see the dog as a capable individual who knows what he wants and how to communicate in order to get it—as someone who has the potential for agency, preferences and choice" (Donaldson and Kymlicka 2011, 84). Does an animal's dependence on human care have to be understood as inevitably negative, as simply characteristics that will invariably equal exploitation? Is it possible to have a relationship with domesticated animals in which humans recognize the value of these animals we have evolved with, beyond a simple calculation of mutual advantage? Could the dependence of domesticated animals be seen as an opportunity for humans and animals to coexist together, as farmers say, "symbiotically," but without accepting exploitation?

A large part of caring for animals ethically means listening to what animals are telling us about the care they are receiving and the care they would like to receive. As Gruen suggests, deciphering what animals need and want is a process that not only demands that we be actively involved in our own empathetic responses to animals, but also that we invest energy into learning about their species-typical behaviors as well as their individual characters. What could we learn from domesticated animals if we took Gruen's suggestions to heart? If we tried harder to listen to them, would it challenge the often infantilized image animal advocates have of animals as "voiceless" beings who simply need our protection? Would our narratives of their futures (as individuals and as species) be altered? As Donovan writes,

1991. Nine-year-old vegan Nellie McKay sends a report to Feminists for Animal Rights about her activism on behalf of animals. In just a few years, McKay's career as songwriter, singer, and performer takes off while continuing to speak on and advocate on behalf of feminism, civil rights, animal rights, and other deeply felt progressive ideals.

"It is not so much … a matter of caring for animals as mothers (human and nonhuman) care for their infants, but of listening to animals, paying emotional attention, taking seriously—caring about—what they are telling us" (Donovan 2006, 305).

Where are we left, though, if both the argument for domesticated animal extinction and the argument for continued slaughter and animal exploitation are both inadequate? I have suggested that to view the dependence of domesticated animals through a disability studies and justice framework may give new answers to the questions surrounding animal exploitation and may also open up a third path in the question of our responsibility to domesticated animals. Instead of continuing to exploit animals because they are dependent on us, and instead of leading these animals to extinction as a potential vegan alternative, could we not realize our responsibilities to these animals whom we have helped create? Could we not recognize our mutual dependence on each other, our mutual vulnerability, and our mutual drive for life? As Donaldson and Kymlicka suggest, could we not recognize that we are all residents—citizens—of shared communities? The big questions that affect disability seem equally relevant to this debate: How can viewing these issues through a lens of interdependence help reframe the conversations? How can those who are seemingly most vulnerable also be recognized as useful, valuable, and necessary? How can we begin to start listening to those who need care about how they feel about their own lives and the care they are receiving?

Feminist disability studies scholar Christine Kelly's description of care seems particularly important here. Kelly explains care as being "an unstable tension among emotions, actions, and values, simultaneously pulled toward both empowerment and coercion." She writes that "Care is a paradox" and that it is a tension among numerous definitions, "none to be disregarded" (Kelly 2013, 790). Kelly provides an analysis of care that insists on messiness and responsibility. Care and needing care are sites that rather than trying to avoid, we need to be radically attentive to.

For better or for worse our co-evolution with domesticated species has created animals whom we are deeply entangled with, both ecologically and emotionally. These animals remind us that we ourselves are a part of "nature," that we cannot just cut ourselves free from other animals. But they also remind us that we are capable of deep coercion and exploitation—that we have too often dominated those whom we deem dependent and vulnerable. Vulnerability and dependence can be unsettling as they are states that require intimacy, empathy, and self-reflection, but they also hold the potential for new ways of being, supporting, and communicating—new ways of creating meaning across differences in species and ability.

Acknowledgments

This article is adapted from a chapter of my book *Beasts of Burden* (forthcoming from the Feminist Press, all rights reserved). I would like to thank the Feminist Press for allowing me to share this version here. This work builds on ideas I first elaborated in two previous articles; I am thus grateful to the *Monthly Review* and *Qui Parle* for allowing me to include some of the content from these pieces. I want to thank Lori Gruen and Carol Adams for their encouragement. And, as always, I am grateful to David Wallace for his love and support.

Notes

1   Someone who is quadriplegic, for example, is not physically autonomous in the same way an able-bodied person is, but this is not necessarily what makes this person dependent. If this person has little to no access to assistant services, to accessible housing, or to transportation, she will at worst spend her life locked away in a nursing home, or at best be at the whim of her family or other volunteer carers, with very little way of changing her situation. However, if this person has access to the social services she needs to live how she desires, to choose and hire who assists her, and an accessible environment in which to live and work, then her life becomes one more of interdependence than of dependence.

2   Allison Aubrey, "Heritage Turkeys: To Save Them, We Must Eat Them," www.npr.org, November 23, 2011 http://www.npr.org/blogs/thesalt/2011/11/23/142703528/heritage-turkeys-to-save-them-we-must-eat-them (accessed October 14, 2013).

3   "Ark of Taste in the USA," *Slow Food USA*, http://www.slowfoodusa.org/ark-of-taste-in-the-usa (accessed October 14, 2013).

4   Jerry Lewis, "What if I had Muscular Dystrophy?," *Parade*, September 2, 1990.

5   Jess Bidgood, "Oxen's Fate Is Embattled as the Abattoir Awaits," (*New York Times*, 2012), http://www.nytimes.com/2012/10/29/us/oxens-possible-slaughter-prompts-fight-in-vermont.html?_r=0 (accessed March 24, 2013).

6   These animal advocates believe that we have a deep responsibility to the animals who exist currently to treat them with compassion and dignity while they are alive. We also have a responsibility to stop

1992. A conference on Women and Animals: Empowerment in the 1990s, sponsored by Friends of Animals, Feminists for Animal Rights, and the feminist magazine, *On the Issues* held in Washington, DC.

breeding billions of these animals every year—after all, there only are so many animals because humans breed them. However, at a certain point a decision will have to be made as to whether remaining domesticated animals are sterilized or kept from breeding to stop their species from reproducing.

7   In regards to disability, I am not arguing that we need to make sure that animals who continue to grow so much muscle mass that their bones break under their weight, or animals whose udders produce so much milk that they are prone to broken bones, infection, and osteoporosis, must continue to exist. Before we can really begin to untangle the massive ethical issues that we have created through breeding, domestication, and exploitation, we have to unpack a lot of complex questions about our responsibilities to different breeds of animals and far more consideration has to go into what disability is in different species and how animals interact with disability. The point is that there needs to be a more thoughtful conversation about disability in regards to domestication and breeding, beyond simply using it as a justification for extinction.

Facing Death and Practicing Grief

Lori Gruen

7

In March 2013, with very little attention, the African black rhino was declared extinct by the organization that is tasked with making such determinations—the International Union for Conservation of Nature. The IUCN also believes the Javan rhino of Vietnam to be extinct after the last known animal there was killed by a poacher in 2010. In between these two little-noticed declarations, in the fall of 2012, an amazing amount of attention focused on the fate of two oxen, Bill and Lou, who worked at Green Mountain College. Lou had injured his leg and students and staff thought Bill wouldn't work without Lou, so after internal discussions the college decided to kill and eat both of them. Tens of thousands of people agitated to have them retired and two sanctuaries agreed to care for them (one of the sanctuaries, VINE sanctuary, is only about an hour away from the college). Despite these offers and after a huge amount of media attention, the college decided to euthanize Lou and quietly dispose of his body. As of this writing, Bill continues to work at the college farm.

Since billions of animals are raised, killed, and eaten every year, this outcry over two oxen may seem somewhat perplexing. Juxtaposed with the fact that so many people failed to notice the demise of two rhinoceros species, the fuss over Bill and Lou seems at best a disproportional response. I find it puzzling that some deaths, or threats of death, generate such attention when others are barely noticed. I will suggest that part of the reason there is such disparity in our responses to death is that those of us who are committed to ending egregious lethal human practices have quite divergent views about what living and dying with other animals entails. These divergent views—coupled with an understandable reluctance to face death—have limited discussions of how we might grieve. In an attempt to open discussions about the construction of compassionate and nurturing practices of grieving as one way to honor the depth of our relationships, I will first analyze divergent views about what it means to live and die with others, and then discuss ways of mourning.

Ecofeminism, extinctionism, and exterminism

Some have argued that we should end our relationships with other animals, particularly domesticated animals, and in that way we can avoid being responsible for their suffering and deaths. These "abolitionists" view our relationships as inherently ones of domination and believe we ought to end them by humanely caring for those animals who are dependent on us and making sure, through sterilization, that no more animals are brought into existence. This view has been called "extinctionism."[1] Many ecofeminists disagree with this view, arguing that attending to both the inevitability and specificity of our relationships with others, humans and nonhumans, and responding with compassion, care, or what I call entangled empathy, are important alternative practices that can lead to changing lives and relationships for the better.[2] Our relationships with other animals (human and non) are a central part of what makes lives meaningful and rather than ending our relationships we'd do better working to improve them.

Ironically, ecofeminists find themselves in tension with some other feminists who mischaracterize ecofeminists as something like extinctionists and who fail to understand the complexities of entangled living and dying. These feminists erroneously see ecofeminists as engaged in some naïve embrace of moral purity and suggest that ecofeminism advocates "exterminism."[3] That ecofeminism is targeted by extinctionists and criticized as exterminist I think reveals an aspect of our complicated relationships with other animals, namely, an anxiety about death and loss that living with other animals inevitably entails.

One of the central insights of ecofeminist theory, an insight that is shared with other theoretical interruptions of the status quo, is that the logic of domination crosses subordinated categories and builds strength from mutually reinforcing exercises of authority. The structures of patriarchy and heteronormativity and racism and colonialism and speciesism are not simply similar or metaphorically connected, but rather, work together (not always or simultaneously) to solidify the power the dominant class gains through the construction of a subordinated "other." Though some get confused by identity politics in debates about these intersections, these are **structures or systems** of power; individuals, even individuals with power within the system, can't create them or undo them on their own and the verdict is still out on what force collective action can have against such powerful systems.[4]

Ecofeminists who are concerned about other animals have been critical of "animal rights" discourse because it is based on a type of individualism that does not provide the theoretical tools for critically examining these systems of exploitation and the conceptual and material forces that help to maintain them. The legalistic reasoning of animal rights also tends to ignore the relationships that we are in and the particular concerns, interests, sympathies, and sensitivities of the individuals in those relationships. In addition, much of the animal rights literature focuses on aggregations and abstractions in two important and sometimes problematic ways. Particular individuals and their relationships are lumped together as mass terms, for example farmed animals, companion animals, research chimpanzees, and suffering, pain, and death are generalized over. This focus on general suffering often excludes an exploration of the other features of one's life that may involve great pleasure or satisfaction in the company of others. Even when most of one's life is lived in conditions of dire exploitation, it is still possible that life isn't all and only suffering, particularly if one has meaningful relationships. Pleasure can sometimes be found in awful places. Ecofeminists have also argued that standard approaches to animal rights truncate the description of the problems that emerge and fail to ask more critical questions about the conditions that allowed systems to create and perpetuate problems in the first place.[5]

While ecofeminists have focused on these systems and on the theoretical differences between approaches to changing the systems that cause suffering and exploitation, we too have not done a good enough job analyzing the particular relationships we have with other animals, particularly companion animals. There is an important concern that living with other animals is characterized by relations that always position humans in control and nonhumans in conditions of servitude or, at best, dependency. Humans are in a relationship of patronage with nonhumans—we can be kind and generous, but always with an air of superiority. When humans bring nonhuman animals into their homes, for example, the nonhuman animals are forced to conform to the human rituals and practices that exist there. Cats and dogs are often denied full expression of their natural urges when their "owners" keep them indoors, in crates, or forbid dogs from digging, scavenging for food, or rolling in the most putrid-smelling stuff they can find. Of course, there are reasons we can give for imposing restrictions on companion animals, some of which have to do with the benefits that are gained by both the animals and their human companions. But even the most thoughtful, compassionate domesticated relationships can't erase the fact that companion animals are forced to live by our cultural standards. Companion animals are, in a very real sense, our captives.[6]

Relationships with companion animals involve degrees of instrumentalization, as perhaps all relationships do, but when we turn to our relationships with farmed animals the power relationship becomes a clear relationship of use. The trouble with these sorts of relationships is most apparent with those animals who are being raised as named "quasi-pets" on so-called sustainable or pasture-based farms. The horrors that accompany the painfully short lives and violent deaths of the animals used for food (clearly on factory farms, but also on smaller farming operations) is what leads many of us to refuse to participate in a system that violently instrumentalizes individuals in deeply troubling ways, obliterates their personalities and interests, and turns them into both real and metaphorical fodder.

Though most of us can readily eschew animal parts in our own diets, ecofeminists are mindful of the violence perpetuated in many gendered, racialized, and colonial contexts as well as the realities of a changing climate and thus forgo top-down, absolute universalizing judgments that everyone, everywhere should see "veganism as a moral baseline." Instead, most ecofeminists argue for "contextual moral veganism" that recognizes both the moral centrality of a vegan diet and contextual exigencies that impede one's ability to live without directly killing or using others (Curtin 1991).

Some theorists have decided that this is not an acceptable response to instrumentalism and instead insist that to avoid participating in exploitation, all of our relations with other (domesticated) animals should end. And the extinctionist view applies not just to farmed animals but also to companion animals. Prominent proponent of this position, Gary Francione, has argued against human relationships with companion animals:

> Domesticated animals such as dogs and cats are vulnerable and entirely dependent on us for all of their needs. They live very unnatural lives because they are not part of the human world and they are not part of the animal world. So however well we treat our nonhuman companions, the institution itself is morally problematic. (Unferth 2011)

I find such views rather arrogant in that they fail to acknowledge the agency of other animals as well as ignoring human vulnerability and dependency. Further, such views suggest that vulnerability and dependency are problematic. Only from a place of privilege might one even formulate the illusion that dependency is something that can be completely overcome. Not only are we in a shared community that would be destroyed if some of us were to be forced out of existence, but others (human and non) co-constitute who we are and how we configure our own identities and agency, our thoughts and desires. We can't make sense of living without others, and that includes other animals. We are entangled in complex relationships and rather than trying to accomplish the impossible by pretending we can disentangle, we would do better to think about how to be more perceptive and more responsive to the deeply entangled relationships we are in. Since we are already, inevitably in relationships, rather than ending them we might try to figure out how to make them better, more meaningful, and more mutually satisfying.

Recognizing that we are inevitably in relationships to other animals, replete with vulnerability, dependency, and even some instrumentalization, and working to understand and improve these relationships is not condoning exploitation. Acknowledging that we are in relationships doesn't mean that all relationships are equally defensible or should stay as they are. Relationships of exploitation or complete instrumentalization are precisely the sorts of relationships that ecofeminists have argued should change. Extinctionists have argued otherwise, suggesting that ecofeminism assumes the legitimacy of institutionalized exploitation and accepts the very hierarchy we ostensibly reject:

> … ecofeminism retains the hierarchy of human over non-human: the nonhuman is treated differently from the human in that the latter is regarded as a "person" while the former is regarded as a "thing." This is explicitly in contrast to rights theory, which eliminates the "thing" status of at least some animals. Ironically, ecofeminism systematically devalues animal interests because it regards the categorical rejection of institutionalized exploitation as itself a hierarchical position. (Francione 1996, 103)

In addition to being a mischaracterization of the ecofeminist position, this claim is too simple and dichotomous. In rejecting universal prohibitions, one needn't accept the value dualism of humans over nonhumans, although navigating the complexities of context in non-hierarchical ways does require care and a good deal of attention.

While extinctionists have argued that ecofeminists condone domination of other animals by refusing to accept universal abolitionism, some feminists have strangely criticized ecofeminists for proposing universal abolition—what they call exterminism. Donna Haraway, for example, attributes this "vexing" view to ecofeminists. As she puts it, exterminism is a sort of killing, a "genocide of ways of living and dying together." It's not clear that Haraway wants every way of living and dying together to stand—stopping actual genocides (think Rwanda, Bosnia, Nazi Germany) denies a certain way of living and dying together but it would be perverse, or maybe ironic, to call ending a genocide a genocide as well. She does think there are ways of living and dying with other animals that it would be extreme or exterminist to prevent. She doesn't clearly lay out what those ways are—it seems that eating a wild pig that a colleague hunted and killed at a department gathering in a part of the world where plant-based foods are readily available should not be condemned, as the activity of roasting a pig reveals our conflictual, cosmopolitical engagement (Haraway 2008, 298–300). Breeding and living with working dogs to prevent such relationships from becoming the stuff of stories is another practice she thinks should be maintained, or at least tolerated (Haraway 2008, 105–6).

Haraway would not make the same argument about cosmopolitical engagements of this sort if the "other" were human, and in this way Francione's criticism of ecofeminism is more aptly directed toward her views. And her criticism of ecofeminists is more aptly directed at his version of vegan abolitionism. Yet ecofeminism becomes the target of both.[7]

Living and dying

Though the charge of exterminism misses its mark when directed at ecofeminists, there is nonetheless something important to be teased out of Haraway's overreach, a kernel of insight that also reveals another problem with extinctionist views. Haraway writes:

> In eating we are most inside the differential relationalities that make us who and what we are ... There is no way to eat and not to kill, no way to eat and not to become with other mortal beings to whom we are accountable, no way to pretend innocence and transcendence or a final peace. Because eating and killing cannot be hygienically separated does not mean that just any way of eating and killing is fine, merely a matter of taste and culture. Multispecies human and nonhuman ways of living and dying are at stake in practices of eating. (Haraway 2008, 295)

We can't live and avoid killing; this is something I think has been underexplored in vegan literature. Of course, vegans kill less than those who aren't attending to other animals or making excuses for their continuation of deathly practices, but the unavoidability of doing harm and causing death even while trying to prevent it demands further reflection.

Living today, even for vegans, involves participating unwittingly in the death of sentient individuals. We can rail against the massive violence that is done to the huge number of living beings who did nothing to deserve their tragic fate, but our political commitments and moral outrage doesn't clean our hands. We harm others (humans and nonhumans) in all aspects of food production. Many are displaced when land is converted for agricultural purposes, including highly endangered animals like orangutans who are coming closer to extinction as a result of the destructive practices used to produce palm oil, a ubiquitous ingredient found in a large number of prepared "vegan" food products. Animals, birds, and insects are killed when fields are tilled and plants are harvested. Though it is hard to calculate the harms to human and other animals from climate changes that result from greenhouse gas emissions from the agricultural sector, it is impossible not to contribute to these harms and still eat. Vegan diets are less harmful than those that include animal products, to be sure, but the harms and deaths occur nonetheless.

Vegans have attended to the tragedy that farmed animals experience, but have generally paid less attention to losses caused by our own practices. The system we live in, work in, play in, and benefit from (or at least make our way in) is built on the backs of other animals. And it is important to get perspective on our complicity in the pain and death of others and think about how we can address that loss. There is no glaring line between "them" and "us," calculating exploiters and pure-of-heart vegans. If we think of these relations as complicated causal networks, we are certainly farther out on the web than those who consume animal bodies, who kill animals directly, and who directly profit from the death of other animals, but we are still in the web, not beyond reproach.

In addition, there has been a growing public perception that vegans see themselves as better than non-vegans, morally superior, preachy, and even annoying. Whether or not that is in fact the case (and there is some evidence that it is probably not as bad as it sometimes looks),[8] it would behoove vegans to reflect further before donning their t-shirts that read "Nobody is perfect, but vegans are close." Considering ethical ways of living amid dying and of the possibility of recognizing inevitable instrumentalization is tricky if we are at the same time figuring out ways to engage in some form of moral repair (Walker 2006). Having a better perspective can only help us to actively and effectively minimize the suffering and death of other beings and make the deaths that do occur meaningful by mourning the lives of those who have died.

1993. Jim Mason's *An Unnatural Order: Why We Are Destroying the Planet and Each Other* published.

Living necessitates dying, and—controversially—killing. We can't live without killing others or, at best, letting them die. When we live with particular cats and dogs, for example, other animals will have to die. Most obviously to feed those animals.[9] Even if they are vegan, dogs and cats will kill and eat other animals if they get a chance. And when we deny them that opportunity, it becomes more obvious how problematic our power over them is. If we are all vegan, growing plants to feed ourselves and other animals involves killing some other animals. Even if some vegans could carefully grow plants in such a way that they don't displace the animals who live on the land they are using, they grow enough to share food with the "denizens" that may raid the fields, and they don't destroy the animals who live in the earth they till, very few can afford to create food in this way.

Not everyone can afford to care for other animals in need and it is simply impossible for those of us who can afford it to provide for all those other animals. In the US, four million dogs and cats are killed every year in shelters. That number is decreasing as more and more people realize that purchasing companion animals when there are already many cats and dogs who need homes is problematic. Nonetheless, approximately 10,000 cats and dogs die every day. No one can continuously adopt more animals and adequately care for those they already live with. As much as I would like to bring more dogs into my life, Maggie and Fuzzy, the dogs I am currently cohabitating with, would not approve. I can see there are reasons, perhaps good reasons, for me not to adopt more dogs, but the fact that I don't means that someone will probably die and I didn't save her. I can't fall into a pit thinking everything is hopeless, but I do think further reflection on the deaths we are responsible for is necessary by members of the dominant species, even vegans. Some unwanted companion animals may end up homeless, some may be warehoused in no-kill shelters, provided with the bare minimum to exist, and those harms may be worse than death. Some no-kill shelters are like storage facilities and animals live in cages with no possibility of being themselves. If the choice is between existence without joy, without relationships, without exercise, without fresh air and sunshine, and non-existence, I think the choice is painfully clear.

And this is another reason to reflect on the deaths we are responsible for rather than avoiding this difficult topic. If companion animals do not have any reasonable hope that they will experience day-to-day pleasures then perhaps it is better, all things considered, for them not to live. When I think of my own death, I can only hope to die with dignity after a really good happy day with the beings I love. Far too often, people make that judgment for companion animals relatively quickly and too easily, and that is shameful. But in some cases it may be our responsibility to make that difficult final decision and we are shirking our responsibility when we don't and pass the buck to others. When we do decide, it is often difficult to admit or find community; we make these decisions in "private" and there is very little wider deliberation about options and endings—this is another problem with not facing death.

Even in our activism we inevitably end up focusing our attention in a way that saves some while countless others die. When the focus is on saving the lives of two named individuals, like Bill and Lou whom I mentioned at the outset, activists are in some sense allowing the tens of thousands of animals killed every day to feed college students to die without fighting as hard for them. Indeed, it is rarely possible to generate enough public commitment to fight the way people did to try to save Bill and Lou, and thus the nameless billions of animals will become someone's food or resource somewhere in the world. There are all sorts of animals who will live and die by the choices we individually and collectively make. Not eating animals doesn't mean vegans aren't involved in killing, although it does mean we are killing less. It's a hard acknowledgment, but rather than avoiding it, perhaps it would be instructive to face it and grieve the lives of those lost while at the same time working to try to minimize causing additional harm and death.

Of course, most vegans are particularly sensitive to the death of others and turning attention to the source of so much sadness can be debilitating. But not acknowledging it is also painful. As James Stanescu has so poignantly written,

> Those of us who value the lives of other animals live in a strange, parallel world to that of other people. Every day we are reminded of the fact that we care for the existence of beings whom other people manage to ignore, to unsee and unhear as if the only traces of the beings' lives are the parts of their bodies rendered into food: flesh transformed into meat. To tear up, or to have trouble functioning, to feel that moment of utter suffocation of being in a hall of death [the meat isle in the grocery store] is something rendered completely socially unintelligible. Most people's response is that we need therapy, or that we can't be sincere. So most of us work hard not to mourn. We refuse mourning in order to function, to get by. But that means most of us, even those of us who are absolutely committed to fighting for animals, regularly have to engage in disavowal. (Stanescu 2012, 268)

That living with other animals and fighting for some other animals also means that other animals will inevitably die, suggests that it is important to come to terms with the death and dying, the grief and mourning that come with being vulnerable, embodied, fragile animals.

1993. Maria Mies and Vandana Shiva's *Ecofeminism* published.

Practicing grief

Urging that we come to terms with death and dying is not an endorsement of the way things are or a retreat to quietism that allows the status quo to continue. But it is an expression of humility and an acknowledgment of the limitations of good intentions. There are genuine moral dilemmas, perhaps fewer than some would think, but there are times in which whatever we do there is some moral "remainder," something that is lost. Acknowledging loss helps us to recognize what Judith Butler has called the "precariousness of life"—a precariousness that is at the core of community. Because we can suffer and be killed, we can recognize relations and kinship. As she writes, "there is no life without the conditions of life that variably sustain life, and those conditions are pervasively social, establishing not the discrete ontology of the person, but rather the interdependency of persons, involving repro-ducible and sustaining social relations, and relations to the environment and to non-human forms of life" (Butler 2009, 19). To recognize our precari-ousness, and our vulnerability, is to not only recognize our interdependence or entanglement, but also, as Stanescu has suggested, to honor our animality.

Living with other animals requires paying more attention to grief, mourning, and maybe shame. This is doubly difficult today as acts of mourning human loss are increasingly sequestered. Sandra Gilbert recounts the ways that both death and grief have been relegated to social margins and illustrates how mourning has been pathologized.

> Given that every healthy mourner is potentially an unhealthy melancholic, it oughtn't be surprising that those who are bereaved feel embarrassment, anxiety, even shame … That neither the mourner herself nor her comforters are able to draw upon any culturally agreed-upon procedures for grieving intensifies her embarrassment even while it may further shape her shame. (Gilbert 2006, 257)

Mourning is also feminized, considered a private activity that must be kept at home. Since keeping grief in the closet and not dwelling on death is an expectation in the human case, even acknowledging the death of animals, let alone mourning those deaths, leads us into the realm of either the comic or the crazy.

But if we follow common norms of grieving and turn away from mourning, from the fact that we are in fatal interactions with other animals, we are not just making their lives unintelligible but are also foreclosing an appreciation of what it means to live with other animals, many of whom have life spans that are shorter than our own. It's not easy to turn our attention to this underside of living with animals; it is painful and it requires courage in speciesist societies, but doing so may allow us to develop new ways of finding meaning in our fleeting relationships. It can also help us in thinking through some of the dilemmas that we currently face in our non-ideal, troubled relationships with animals. This is especially important for animal advocates who work on the front lines of animal care in sanctuaries and shelters.[10]

The practices that lead to the suffering and death of other animals are practices that, in addition to causing pain and ending life, also render those lives meaningless. Developing counter-practices of mourning can help make those lives and, importantly, our relationships to those who are now gone, intelligible. But since there are few recognizable practices for grieving humans in contemporary society, developing meaningful mourning practices for other-than-human animals is all the more challenging. Some animals appear to have developed what we might think of as mourning rituals and perhaps we can learn something from them.

There have been many reports of dolphin, ape, and monkey mothers who carry around their dead children. Katie Cronin video-recorded the reaction of Masya, an 18-year-old wild-born chimpanzee who had been living at Chimfunshi Wildlife Orphanage Trust in Zambia for 15 years, to the death of her infant daughter. Masya had reared two other daughters, one of whom, Mary, was in the social group at the time of the infant's death. For two days Masya interacted in various ways with the corpse of her infant. The first day she held on to her and carried her around, a behavior that has been observed in captivity and the wild in many different species. The second day, Masya put the dead infant down, but then repeatedly would check on her (Cronin et al. 2011). Reports from other observers indicate that some mothers will carry their infants much longer.

Cynthia Moss and others have described the mourning rituals of elephants. Elephants pay homage to those that die and often walk miles to return to the site where a member of their family has passed. In some instances, members of other family groups come to pay their respects. And elephants will handle the bones of members of their family long after they have died. Echo, the remarkable matriarch that Moss observed for almost 40 years at Amboseli, lost her daughter Erin, who died from an infection caused by a spear wound. Erin had a young son and as she was dying Echo took him away to get food and water and presumably so he would not witness his mother's death. But the family later returned to touch and feel Erin's bones. Researchers have noted that "African elephants are unusual in that they not only give dramatic reactions to the dead bodies of other elephants, but are also reported to systematically investigate elephant bones and tusks that they encounter" (McComb et al. 2006, 26). They will also pick up tusks and carry them around in their trunks.

1993. The Bloodroot Collective's *Perennial Political Palate: The Third Feminist-Vegetarian Cookbook* published.

Deborah Bird Rose shares a multi-species experience of grief in which a pair of humans and a pair of albatross mourn the loss of the albatross's egg. Male and female albatross take turns sitting on the egg, while the other is off feeding. After the male had been sitting on the egg for five weeks and the female did not return, he grew weak and had to eat. Louise and Rick, the human couple, were worried about what would happen to the egg and were relieved when both the male and female returned. But it was clear that things were not well. Here is how the humans describe what happened when they noticed that the egg was broken:

> … it was really sad. We did nothing but cry the whole day, pretty much. Because they, Makana and Kupa'a were out there mourning and crying … we were all crying. You could tell it was a different sound. They were doing the "sky moo," but instead of their "oooh, oooh" it was "aah, aah." It was sad. Awful. Just awful … she kept on trying to sit on it, and he would talk to her. He was starting to groom her. And she started to appear to realize that there was a problem with the egg, and they started to grieve. She really struggled to accept it—the loss of their chick. (Rose 2013, 7)

According to Rose's report, Louise and Rick acknowledged that they could only explain what they were experiencing in anthropocentric terms and admitted that "even if their interpretation was wrong" there was an obvious change in behavior. Together Louise and Rick and Makana and Kupa'a experienced a loss that they all had to come to accept.

It's hard to know exactly whether Masya, or Echo, or Makana and Kupa'a or the others were really grieving. Like Louise and Rick, we need to be cautious in our analyses of these different behaviors so as to avoid simply projecting our anthropocentric interpretations on their unique practices. And it is important to attend to the diversity of experiences of grief. Some chimpanzees, dogs, or hens appear to mourn the loss of a group member, while others seem not to care that much. Human grief practices are also quite different from those of other animals, but this doesn't suggest that human practices are necessarily deeper or more significant. As Barbara King notes in her book *How Animals Grieve:*

> … they aren't the story tellers we are, passing down elaborate narratives about our grandparents and parents to our children and grandchildren. Does that mean our grief is deeper than the grief of chimpanzees? Questions like this one miss the point. We each are what we are, animals bound together by our various ways of grieving. (King 2013, 148)

Whatever grieving behaviors might mean to those who mourn, we can learn from them in developing our own mourning practices. When my feline companion of 23 years, Recatsner, died, I had her body cremated and in the remains I discovered parts of bone and I had the small bone made into a pendant. It turns out there is a whole (apparently quite lucrative) cottage industry making "pet" cremation jewelry. While my desire to carry a part of Recatsner with me is satisfied, this is not a communal mourning practice. I am not connected to others who are grieving when I wear the pendant; indeed, unless I told someone, no one would see it as a memorial.

Creating communal possibilities for mourning our companions whom we have loved and lost as well as all of the other animals for whom we grieve can take grief out of the closet, out of the realm of the comic or crazy, and make the lives and deaths of other animals visible and meaningful. There are models of public grieving in protest, ACT UP and anti-war protests turn private losses into collective mourning. The recent campaign started by Israeli activists to have the number 269 permanently marked on bodies (through branding or tattoos) to stand in solidarity with a calf who has a tag with the number 269 through his ear, although not conceived as a campaign of community grieving, could be. Identifying with calf 269 and having 269 permanently marked on one's body in protest of the lethal practices that destroy the lives of billions of animals may also be thought of as a form of mourning. It memorializes and makes explicit the individual whose life is lost and puts those who have marked themselves in solidarity into community with one another.

While so many suffer and die, and while their deaths pain us, it is vital that our relations to other animals and each other not be exclusively mournful. When there is a loss, there has to be a way to emotionally and physically absorb and process—to metabolize—the loss. In the case of animals who suffer and die in factory farms and laboratories, as well as those who are dying as ice caps melt and habitats are destroyed, we need to metabolize communally. Perhaps we might develop a ritual vegan feasting practice, to share in our grief, to memorialize and mourn those who have died. Collectively grieving provides a way to honor the precariousness and fragility of our entangled lives.

Acknowledgments

I began thinking about these issues in response to a powerful presentation by James Stanescu at the Sex/Gender/Species conference that I co-hosted at Wesleyan in 2011 and I thank Scu for prompting my ongoing reflections on grief and mourning. The death of Marti Kheel and the support that ecofeminists provided each other during our grieving specifically pushed my thinking forward, thanks especially to Carol Adams for her nurturance when we were mourning our loss of Marti. I also want to thank Greta Gaard and pattrice jones for helpful reflections on an earlier draft of this chapter and express my deep gratitude to Maggie, always. I dedicate this chapter to Fuzzy—whose joy in life, boundless love, and heartbreaking death as this book was going to press, provide tremendous inspiration.

Notes

1   Sue Donaldson and Will Kymlicka (2011) describe the extinctionist view as one that "seeks the abolition of relationship between humans and domesticated animals, and since domesticated animals can rarely survive on their own, this is effect means the extinction of domesticated species ... According to the abolitionist/extinctionist view, the horrendous history of injustice leads to an inescapable conclusion: we must remove ourselves from the equation—whether as owners, overlords, stewards, or ostensible co-contractors ... we cannot have domestication without mistreatment, because mistreatment is intrinsic to the very concept of domestication" (77–8). Donaldson and Kymlicka reject this view.

2   Not all ecofeminists agree here. Since the 1990s, there have been debates between "vegetarian ecofeminists," sometimes called "animal ecofeminists," and other ecofeminists about attention to the lives and deaths of other than human animals. See Gaard 2002, for example.

3   Donna Haraway uses "exterminism" in two different senses in *When Species Meet*. She notes the sense that Derrida uses to describe the horror of "immense, systematized violence against animals," but later she broadens the logic of sacrifice to recognize not just that "it is not killing that gets us into exterminism, but making beings killable" (80) but that "moral absolutes contribute to what I mean by exterminism" (106). For a robust critique of Haraway, see Weisberg 2009.

4   That is not to say that we shouldn't collectively work to uncover and undermine these systems; it is simply to recognize the complex forces of such systems and the difficulties involved in dismantling them.

5   For more extensive criticisms of animal rights discourse see Donovan and Adams 2007, Kheel 2008, Gruen and Weil 2012.

6   For a discussion of some of the issues of captivity of cats, dogs, and other animals see Gruen 2014.

7   There is an important question about why ecofeminism has become the target and more work can be done on figuring out why ecofeminism

continues to be mischaracterized in these ways. While I think anxiety about death, as I've suggested here, is one answer, there are probably others.

8 For example, Tania Lombrozo's 13.7 NPR blog on 11/26/2012 notes: "Vegetarianism is a blossoming field of study, with research in psychology and other disciplines exploring the characteristics of vegetarians and omnivores, as well as people's *perceptions* of vegetarians and omnivores." http://www.npr. org/blogs/13.7/2012/11/26/165736028/its-time-to-end-the-turkey-tofurky-thanksgiving-food-fight. Based on a Public Policy Poll, as reported by the Humane Research Council, "Among American voters, 49% view vegetarians favorably, and 22% unfavorably. In a pattern that holds for most cross-tabulations, vegans are seen less favorably by about 10%, yielding 38% favorable versus 30% unfavorable. http://www.humanespot.org/content/who-views-vegetarians-vegans-positively-new-poll-results

9 There is still a great deal of controversy about the nutritional adequacy of vegan diets for dogs and especially for cats.

10 Perhaps more than anybody else, sanctuary care-givers have to deal with death and dying and loss and grieving often while at the same time providing care for others and so may not have time to acknowledge their grief. Fortunately, some sanctuaries have created memorial parks or areas where staff and visitors can mourn. Announcements in the form of obituaries and personal reflections are sometimes published. But grieving is still quite an individual activity and public practices of grieving, when they occur, seem to take the form of Facebook comments.

1993: Cesar Chavez, co-founder of United Farm Workers and civil and animal rights activist, dies.

part two

Context

Figure 8.1 In what contexts is omnivory celebrated and in which is it to be forsaken? Photograph by Kenneth Allen of sculpture "The Rites of Dionsysus" by Tom Shaw (Eden Project, Cornwall UK) (Creative Commons Attribution-ShareAlike 2.0 license).

Caring Cannibals and Human(e) Farming: Testing Contextual Edibility for Speciesism

Ralph Acampora

8

Imagine the following scenario: by some stroke of historical, geographic, or extra-terrestrial displacement, you come into contact with a lost tribe of foragers whose living and eating customs betray an odd mix of the benign and the exploitive. These folk—who appear in most ways human, if not exactly humane—seem for the most part to be fairly decent members of the life community. And indeed their culture is rather advanced, replete with a wide spectrum of practices and achievements in narrative, technics, dance, music, and science. Likewise, they have become quite refined in their level of ethical discourse and moral development, and can often be found discussing various theories and controversies regarding the right manner of conduct and standards of good character.

One thing horrifies you, however: this lost tribe occasionally preys upon other humans—not members of their own ethnic

1993. Delora Wisemoon begins in Austin, Texas a program for sheltering the companion animals of battered women.

community, to be sure, but rather wandering loners and those from tribes and nations whose territory lies adjacent to or somewhat overlaps the country inhabited by the lost tribalists. Hunts are conducted, killing consummated, and human flesh devoured. In a certain light, it is true, there can be no blaming the lost tribe—for long ago, nobody remembers when or why anymore, they came to reside in the remotest recess of land habitable by hominids; there vegetation is sparse and even other animal life is only rarely found, with the result that nutritional needs must be met by additional "supplements."

What's more, it should be noted with some admiration that the lost tribe has evolved elaborate rituals, mythic lore, and even a kind of sympathetic spirituality that centers on their marvel and respect for the other people they eat. The tribe itself constitutes its identity through a braiding of totemistic allegiances with the more predaceous species to be found elsewhere around the continent of its residence. There are those who identify with the tigers or lions, with the bears or wolves, with the hawks or raptors, with the orcas or sharks—and these identifications are themselves subtly undergirded by attachment to the various kinds of elemental medium and ecological niche available to the heterotrophic organisms occupying this quadrant and epoch of the (or some alternate) biosphere. Now many of these totems' actual referents are far removed in the present distribution of species—but there remain rich oral and literate traditions linking the clans of the lost tribe back to an imagined evolutionary consanguinity that they take to be "realissimo," or the true reality of their common essence, before the migratory crosspaths of Diaspora flung animal kin far and wide.

Consequently, by means of complex psychological mechanisms and religious liturgies, when the lost tribalists hunt other humans they enter into a trance-like state of virtual communion with their prey—re-enacting, in effect, the thrilling chases of yore between select totem species and ancestral hominids. Before setting out, they recite venatic scripture and chant songs of eulogy to the great Web of Life and Death. The successful hunter is taken to be the one who consummates the kill with no cruelty at heart and the least possible infliction of pain or suffering—she claims experiential singularity with her kill, and is granted a key role at the Rite of Ingestion (which is duly supervised by elders no longer of sufficient physical vigor to conduct their own hunts). The central moment of this ritual features a prayer of forgiveness to and appreciation for the sacrificed specimen.

After this gesture of atonement, an entire clan may join the feast and at different stages of participation there are readings from their Great Book of Transfiguration, texts that testify to the wisdom encoded in their esoteric Doctrine of Metabolism. This body of dogma and the associated commentaries and teachings based upon it suggest that the cannibalized prey effectively offered itself for life-merging with the hunter *qua* totemized predator. Thus we find a profound paradox that runs through the conceptual and affective dimensions of the lost tribe's shocking practice of "caring cannibalism"—namely, that they appear to think of this tradition as justified, on the one hand, whereas on the other they feel (perhaps subconsciously, and yet sufficiently) shameful to offer ritualistic apologies for the killing and eating of other humans. Still, in virtue of a mystical intuition that resists articulation, they somehow resolve this tension at a level of experience satisfactory for continuance of the custom.[1] Indeed, so successful is this ineffable resolution of seeming incoherence that the proudest phase of passage for a juvenile member of the lost tribe (and a great pleasure to his family) is observed to be the sacrament of First Consumption.

What are we to make of this hypothetical scenario? Are the "caring cannibals" above a lost tribe merely in the sense of spatial or temporal remoteness—or have they lost their moral bearings as well? Is the putative parable just a far-fetched puzzle that exercises certain (kinds of) philosophers who happen to be fond of so-called thought experiments? Perhaps. Yet maybe the story just sketched can serve as a heuristic analogy that could open up some fertile ground for reflection upon contextual moral vegetarianism (CMV). For our present purposes, let us consider it from the perspectives of some exemplary ecofeminists who theorize on the basis of, or at least in concert with, care ethics. Karen Warren, for example, concludes that "the bottom line for a [care-sensitive] contextual moral vegetarianism is that food practices regarding animals should not grow out of, reflect, or perpetuate oppressive conceptual frameworks and the behaviors such frameworks sanction" (2000, 143). Taking this assertion as a criterial proposition about im/moral edibility, whereby those practices that do uphold oppression are (to be deemed) unethical and those practices that do not are (to be deemed) ethically permissible, it would seem that the caring cannibals' consumption of certain human animals meets the test and thus is an allowable food practice. That would be the result if we took the standard's conjunction strictly—namely, that food practices should not depend on or promote *both* conceptual *and* behavioral oppression (whereas caring cannibalism is associated *only* with behavioral but *not* with conceptual oppression). More charitably, however, if we took the standard as a severable, two-part means of judgment, then we could say that the cannibals' behavior is contemptible and yet their caring conception is laudable—because it can wear an egalitarian face that does not trade consciously on the "Up-Down hierarchical thinking" (139–40) excoriated by Warren (i.e. it needn't be accompanied by a supremacist ideology). Thus, if we share the common intuition that human cannibalism is outrageously

1993. First Open Rescue by Animal Liberation Victoria, Australia.

wrong, Warren's CMV can be reduced to absurdity (totally on the first construal above, partially on the second).

Another way to register this reduction is to view the hypothetical scenario through the lenses of Warren's critique of "the logic of domination."[2] Succinctly put, her rendition of ecofeminism has it that any ecofeminist philosophy worth its salt will identify and reject all instances of this theoretical and practical matrix. The logic of domination, as she analyzes it, consists of four main gestures. First, it confronts otherness not by taking divergences of difference at face value, but instead by dualizing alterity into the polar opposite of some already familiar quality. Second, it configures these dipoles along a metaphorically vertical axis of hierarchy, such that one (self/same/familiar) is thought of as "above" the other (stranger or dualized difference), which is thus conceived to be "below." Third, the logic of domination invests these positions with axiological significance rhetorically capable of legitimating exploitation: "up" entities are "better" than "down" ones, and that means the Downs may be justly used as means to ends of correlative Ups. Fourth, and finally, this rationalization motivates and reinforces the implementation of exploitative practices executed by Ups upon Downs (which can be called oppression correctly, if but only if the exploited is one capable of having choices or options and at least some of those choices or options are in fact squelched by the exploiter).

Judging the moral turpitude of the lost tribe, as I have described it above, from the perspective of challenging domination's logic, as Warren conceives of it, becomes a difficult exercise that decidedly does *not* yield ethical clarity. While it is undeniable that the caring cannibals are oppressive exploiters of their human prey, it is by no means clear to what extent they participate in the logic of domination. Most telling is that they may very well conceive of those people they eat as equals or peers, rather than subhuman underlings: the eaten are different, to be sure, in that they are made edible by the eaters—but this maneuver need not entail polarizing the difference, nor does it necessarily imply constituting that difference as inferiority. Here the caring cannibals can be likened to the Amazonian Guaja, who conduct symbolic cannibalism of monkeys with whom they identify: the practice is apparently underwritten by a cosmology that acknowledges "kinship between all living things" and believes that "like eats like"—such that "for the Guaja, it is the underlying and observable similarities between humans and monkeys which render monkey edible" (Hurn 2012, 94). Now, granted the extreme similarity (taxonomic identity) between the caring cannibals and their own prey, we may wonder whether a Guajan-type principle could sanction actual (rather than merely symbolic) cannibalism. Even if not, the lost tribe does feel justified in its cannibalistic exploitation, but in their view the practice's justification comes from (perceived) necessity—*not* from a hierarchy of being or value that legitimates oppression. In this respect they believe themselves to be operating not unlike how certain naval laws of ours excuse cannibalism in cases of shipwrecked or marooned exigency. Thus it would seem that the caring cannibals' morality need not rely, in any heavy sense, upon the logic of domination targeted by Warren's brand of ecofeminism.

Suppose now that we switch our viewpoint from Warren's to that of Val Plumwood, who rejects global vegetarianism on the basis that not all use of animal flesh as food is necessarily disrespectful or uncaring. Her stance is informed by the rather intense and nearly unique experience of having survived an Australian crocodile's death-rolls, which taught her what it means and feels like to be(come) prey for another being (1995).[3] This experience results in her rejection of "ontological vegetarianism" that "demonizes human (and animal) predation and predation identities" (2000). In this context, Plumwood makes an analogy between heterotrophy and human sexuality thus: "saying that ontologising earth others as edible is responsible for their degraded treatment as 'meat' is much like saying that ontologising human others as sexual beings is responsible for rape or sexual abuse" (2000). The point here is that both claims are errant in that they confuse necessary for sufficient conditions in their respective cases. Of course, it is fair to point out a relevant disanalogy between almost all carnivory and benign sexuality—namely, that the latter typically involves mutual consent while the former does not. Against the rejoinder that consensuality simply does not apply to non-linguistic entities, it could be observed that prey animals refuse an interpretation of willingness by their very efforts of avoidance. Finally, the universalist vegetarian may distinguish between instinct-driven predation of and by nonhuman animals versus the responsibility for diet that attends the moral agency of mature human personhood.

Another tack that might be taken on behalf of CMV à la Plumwood would be to claim that putative or actual instances or practices of flesh-eating can be counted as respectful/caring usage by meeting the test of Kant's categorical imperative (in its teleological formula), because they would *not* amount to some moral agent (the eater) treating some moral patient (the eaten) *merely* as a means to the former's end (of nutrition, e.g.).[4] Hence such flesh-eating might avoid critique from deontological perspectives compatible with the rationale undergirding certain (e.g. Tom Regan's) animal rights advocacy—because "what is prohibited is *unconstrained or total* use of others as means, reducing others [*entirely*] to means" (Plumwood 2002, 159, *italics added*) and not any use whatsoever (a prohibition which would be self-stultifying and arguably suicidal in ecological context). "We cannot give up using one

1994. Carol J. Adams's *Neither Man nor Beast* published, which includes an essay on white privilege, race, and animal issues.

another," concludes Plumwood (2002, 159), "but we can give up use/respect dualism [mutual exclusion], which means working toward ethical, respectful and highly constrained forms of use."

Now the caring cannibals' usage of humans as food can indeed be characterized as respectful (in light of their honorific prayers, valorizing rituals, etc.) and perhaps even as constrained (if it were to happen rarely and/or in low volume). Surely, however, it matters not to the individual being eaten whether the carnivorous practice to which s/he succumbs is limited in frequency or number—and any proponent of universal human rights would object that the prayers and liturgical gestures don't actually benefit the one being killed and eaten; nor, we can observe, would they matter in the case of nonhuman animals—and so, on pain of speciesism, we should not judge any flesh-eating as respectful or morally permissible (at least not *prima facie*). This conclusion could be confirmed by pointing out that eating is literally a case of consumption, of *using up* some-body—such that there is no remainder whom one could any longer respect or care for/about. In cases of other animals at least, Plumwood (as per Warren 2000, 138) is ready to rejoin that respectful carnivory expresses or implies a promise that the eater will reciprocate in due course (i.e. offer her body, upon death, to the food web or energy circuit of the ambient ecosystem); and, knowing Plumwood's feisty attitude and hardcore convictions, it is plausible that she would have been prepared to bite the bullet on this score regarding the caring cannibals (i.e. aver that their cannibalism is justifiable so long as they make a similar commitment of recompense). My point at this juncture is twofold: first, most of us would not be eager to follow suit; second, even if some of us were to join ranks with Plumwood here, there is an eco-ethical category mistake in the notion that a predatory eater's later recycling through food chains or energy circuits—which are *networks or systems*—performs a reciprocation to the erstwhile prey *as an individual* (because the latter is a different sort of being from the former, and moreover it is one who would no longer exist on the very supposition at stake!).

So far, then, we have not found a version of CMV that would (fully) solve the problem of speciesism posed by the story of the lost tribe. Yet there are two more renditions to consider, and they do fare better—not because they solve the issue head-on, but rather because one (Deane Curtin's) **dis**solves it and the other (Marti Kheel's) **re**solves it. If one reads the former's position on CMV (1992, esp. section 4), it at first appears that Curtin would fall into the trap set by my thought experiment: he approvingly cites Wendell Berry's apology for carnivory when the flesh is "eat[en] with understanding and with gratitude"; he underscores that "to live is to commit violence"; and he allows that "some cultures, cultures that provide spiritual self-definition through food, have cultural rituals that mediate the moral burden of killing and inflicting pain for food" (130ff.). All this makes it seem that he would have to sanction the caring cannibals' homicidal eating habit. However, Curtin also writes, "I do not mean to imply ... that the choice to treat other animals as food is morally justifiable in exotic cultures, or in the 'Third World', or in extreme contexts" (132). How can he both uphold cultural pluralism (evident in the earlier references) and yet retreat from approbation of meat-eating in the contexts just cited (as well as, presumably, in my thought experiment)? It is not all that clear—but reading between his lines, I believe the key to grasping what Curtin calls "authentic presence to food" is that heterotrophic living's inevitable "violence needs to be understood within a particular cultural narrative" (132); he could then say that my story is a truncated narrative which would prove impossible if we tried to fill in more details (e.g. how the hominid prey in this remote region putatively subsist themselves) and that such carnivorous narratives as do exist in exotic cultures and extreme contexts may explain yet do not justify their subject matter (or, to put a Wittgensteinian spin on it, that these cultures and contexts present forms of life or exigencies of survival in which the language game of moral justification/condemnation is not played because it simply does not fit). Thus the problem under discussion evaporates—because *neither* eating other animals *nor* human cannibalism is deemed moral.

Although this putative dissolution works well enough on its own terms, it can leave one dissatisfied in that it dodges the issue at stake instead of confronting it. We want to know how and to what extent multicultural tolerance for dietary diversity can coexist with animal advocacy's call for ethical vegetarianism. Here Curtin himself provides clues that lead us to consideration, finally, of Marti Kheel's work. While he admits that "authentic presence to food may express itself differently in different [cultural/narrative] 'worlds'," nonetheless he finds inauthenticity "in western, industrialized countries [where] flesh foods are almost exclusively encountered in contexts that express alienation from and dominance over other beings" (132). Thus, for those of us inhabiting such socio-historical locations, Curtin endorses CMV as "a bodily commitment to direct one's defining relations toward nonviolence whenever possible ... [which] should be understood not as a moral state [of purity], therefore, but as a [meliorist] moral direction" (131). This idea of veg(etari)anism as an aspirational vector is taken up by Marti Kheel in her delicately balanced discourse on the topic. Her version of CMV is animated by a certain sort of universalizing impulse. She was keen to eschew, it is true, the kind of moralism that trades in ethnocentric judgmentalism. Nonetheless, she also held onto the notion of indefinitely widespread veganism *as a regulative ideal* for anybody and everybody interested in what she called "nature ethics" (2008, 235–6).

Following Kheel along this tightrope, I wonder, could a nature ethicist disapprove of the caring cannibals' food practice *and* maintain her CMV *without* the stain of speciesism? It seems to me that a key move here would be to heed the rhetorical mode in which veg(etari)anism is invoked. For Kheel, it is clear that moral philosophy should not be in the business of issuing injunctions that seek to compel conduct or impose belief; rather, her brand of nature ethics pursues an edifying discourse of encouragement where invitation and inspiration are the watchwords (14 and 266n.154). "There is nothing inherently oppressive in encouraging vegetarianism or veganism as ideals," she pleads, "while recognizing that there may be environmental and climatic factors that make them difficult in some cultures" (236). It would appear, then, that nature ethics à la Kheel could consistently advocate herbivory to my hypothetical lost tribe (as well as to actual peoples who may practice subsistence hunting of other animals) and yet maintain its contextual *bona fides*—because, while the ideal of veganism is universalizable in spirit or principle, the manner of address through which it is pursued practically is sensitive to ecological/geographic and sociohistorical conditions.[5] This means that such advocacy is probably best (most respectfully and most effectively) offered to, and then conducted by, critically minded or dissenting members of the relevant groups themselves; and conversely, "[s]ince the vast majority of animal abuse occurs on factory farms that are owned and operated in the Western, developed world, it is most appropriate for vegetarian advocates living in the West to direct their central criticisms to this form of abuse" (267n.161).

This last point of Kheel's raises the issue of agricultural ethics, and it will be helpful now for us to consider such at greater length. For we may well wonder whether a CMV-inflected proscription against factory farming actually invites or at least makes allowance for so-called humane farming practices. Although I am not aware of any self-identified CMV advocate who explicitly endorses animal agriculture of the small-scale, free-range, organic, or otherwise welfarist variety,[6] one can certainly conceive how CMV might be pressed into its service: the farmed animals are taken care of properly (with veterinary attention for *their* well-being, not just for the sake of consumers' health or safety), their species/breed-typical interests are addressed (with decent living conditions assured), environmental and labor concerns are met (through sustainable rearing methods, fair wages, and slaughter techniques duly approved by Temple Grandin)—taken together, all these scruples *prima facie* compose a context in which vegetarianism is optional rather than obligatory, in which omnivory may be indulged without qualm. All things considered, however, most CMV proponents would likely borrow from Warren's critique of the logic of domination: the moral problem with humane farming, it could be argued, is that it relies upon a system of animal husbandry that remains despotic and prone to oppressive backsliding as it perpetuates speciesist hierarchy in a crypto-sexist vein.

Is it possible for humane agriculture to resist the charge of domineering speciesism? One gambit would be for its advocates to bite the bullet on the hypothesis of *human* farming—as, for example, was done in Jonathan Swift's "A Modest Proposal" (1729). After rehearsing the difficulties of poverty and famine confronting an overpopulated Ireland, Swift observed "that a young healthy child well nursed, is, at a year old, a most delicious nourishing and wholesome food, whether stewed, roasted, baked, or boiled" and suggested that a sustainable harvest of human flesh might be instituted by setting aside some Irish children "reserved for breed" while most could "be offered in sale" for consumption and by-products (since it is possible to "flea the carcass; the skin of which, artificially dressed, will make admirable gloves for ladies, and summer boots for fine gentlemen"). Of course the proposal at stake is notoriously satirical (a wicked parody of what could be called flat-footed or callous consequentialism), and it is not hard to discern in its author's persona another kind of pernicious hierarchy at work—namely, an ethnocentrist rationalization of Celto-Gaelic exploitation at the hands and mouths of Anglo-Saxon overlords (thus Swift's references to reducing the population of papist Catholics in order to protect "good Protestant" upkeep of the United Kingdom). So this stab at Malthusian meat-eating (as it were) seems to fail as a genuine legitimation of humane farming, at least if the terms of such are to be cast in the concepts and criteria of CMV.

What would seem needed for a CMV stamp of approval might be a system wherein people harvested their own peers—something like the society depicted in *Logan's Run* (Anderson 1976; Noland and Johnson 1967), with the twist that those culled would also be consumed. Recall that, in this piece of science fiction, conditions of scarcity in the future give rise to a policy of exterminating those above a certain age of maturity (twenty-one in the novel, thirty in the film). Execution is most often accepted fatally—in both senses of that word—by those who reach their allotted limit, perhaps because it is carried out painlessly and valorized by rites of sacrifice. Yet some rebellious spirits do reject the practice and seek "sanctuary" to elude their early demise, which raises the issue of ageism as a form of illegitimate discrimination and gives the narrative a distinctively dystopian atmosphere. One could imagine, of course, that the lethal enforcement of a dispropor- tionately youthful society would be defended by pointing out that there weren't enough resources for a full-spectrum population and that those who

1994. Feminists for Animal Rights publishes two responses to the PETA's "'I'd rather go naked than wear fur" campaign.

reach some age duly calculated for demographic equilibrium have enjoyed whatever period of time is sustainable, whereas those younger than them require and deserve the same opportunity. Still, in both the cinematic and literary versions, the "runners" who attempt to escape their appointed fate take on a heroic aura that threatens to subvert acquiescence to a thanatology of social sustainability. Hence, to the extent that fleeing becomes preferable to dying on time, the maintenance dynamic of a Logan-like society would approximate hunting more than farming—and thus our discussion would revert to the considerations examined above (i.e. those pertaining to the thought experiment about caring cannibals).

Yet there is another work of science fiction that more approximates a model of human(e) farming, namely the film *Soylent Green* (Fleischer 1973).[7] Set at the turn of the millennium (original book), or shortly thereafter (movie), we are again faced with conditions of overpopulation and resource scarcity. Under these circumstances, American food producers apparently try first to manufacture and mass-distribute a vegetarian loaf of concentrated **soy**a and **lent**ils, then move into the market for plankton pies, and end up scraping the bottom of agriculture's barrel by converting human corpses into cakes that resemble and are advertised as nonhuman foodstuffs. Though the details are kept murky, it would seem that the harvesting process occupies a twilight zone between farming and foraging: it is hinted that the supply of human bodies is sourced from natural deaths and from a brisk business of voluntary euthanasia or assisted suicide (some folks find the future bleak and dismal enough to opt out ahead of time). Inasmuch as this industrialized form of cannibalism goes beyond the happenstance of pure scavenging and actually facilitates death upon demand, it can be seen as a kind of agriculture—and one that certain utilitarians might have to accept, if pressed hard enough. But would proponents of CMV have to follow suit? No, at least not on Curtin's version—because it is anchored in the notion of "authentic presence to food" and the marketing of "soylent green" is a real whopper (pun intended, without apology to Burger King)! Likewise, by assiduously investigating inauthenticity in operations billed as "cruelty-free,"[8] an advocate of CMV can also avoid endorsing so-called humane farming of nonhumans.

Taking stock of our reflections to this point, it would seem that at least certain versions of CMV can avoid speciesism—albeit with sometimes fancy theoretical footwork—even in the face of thought-experimental challenges centering on cannibalism of one variety or another. So ... has our discussion amounted to much ado about, if not nothing, then very little? In reply, I remind the reader that there is always value in testing and thereby fortifying a viewpoint. Still, I am somewhat anxiously aware that I may have been testing not only CMV but also my audience's patience with what might have seemed a series of far-fetched fantasies whose critical bite traded on abusing a peculiarly profound taboo. After all, it does remain a conceivable option to avoid speciesism not by enforcing limits on edibility but rather by eroding such boundaries: break the spell of the taboo at stake and accept cannibalism under certain conditions of care or respect—thence argue on behalf of subsistence hunting and humane farming. But does anybody actually take up such a position? As it turns out, there is indeed a theorist in animal studies who has recently given the appearance of doing something very much like this …

In an essay entitled "Cannibalism, Consumption, and Kinship in Animal Studies" (2011), Analia Villagra notes that certain indigenous conceptions of edibility attribute to eating "… the transformational power to make another [the eaten] sacred'" (47), and declares that she "would like to argue for a more challenging vision of kinship that would allow for the consumption of fellow animals not in the absence of or in spite of bonds of kinship, but rather because of them" (50). In a bold theoretic gambit, Villagra argues the case for "becoming cannibal" (52ff.). First, "[t]he reinsertion of human sociality into the wider natural world requires the employment of more animal terms of relation." Thus "we must confront the idea that as the kin of other animals, when we consume them we may become the cannibals we have so feared." Yet "'[t]aking cannibalism seriously' entails releasing ourselves from the cultural mythology of the ruthless, inhuman killer (the Hannibal Lecter mythos) and acknowledging the profoundly transformative aspect of cannibalism."[9] Indeed,

> [t]here is an honesty in accepting a cannibal identity, and cannibal consumption is certainly more productive and noble than the way we currently consume animals … as commodified pieces of factory-farmed meat in a global capitalist marketplace, [for] cannibal consumption seeks to incorporate rather than alienate … [and so] may come to represent a union between two bodies rather than expressing domination of one over the other.

In a final flourish that waxes almost Swiftian, and consistent with the argument just given, our intrepid author advocates eating the companion animal in one's midst as "my delicious pet," because "[c]onsumed kin … are made more kinlike as they are fluidly incorporated into literal [metabolic] and metaphoric acts of cannibalism." Sensationally suggesting that "we move into the uneasy realm of the cannibal as a way to seriously explore the way kinship intersects with and conjoins consumption," Villagra concludes with a dare: "Rather than giving in to the gut [!] sense of disgust that the suggestion of cannibalism may invoke, we can embrace the cannibal, a figure

1995. Animals and Women: Feminist Theoretical Explorations edited by Carol J. Adams and Josephine Donovan.

1995. Carol J. Adams's "Woman-Battering and Harm to Animals" published in Animals and Women: Feminist Theoretical Explorations.

who resolves many of the tensions created by seeing animals as subjects, as kin, and as food."

Because it confronts the taboo against cannibalism head-on, in what seems a fearless fashion, Villagra's essay is rhetorically entrancing and it can have the effect of stifling objections that might appear too small-minded or timid. Nonetheless, it is important to underscore what the essay does *not* do—namely, *it never commends **intra**-specific cannibalism of humans by humans.* No, Villagra stays clear of that practice and remains content to laud cannibalism only when it figures symbolically or else preys upon extended kin in the **non**-human fold. Could it be that the taboo at stake is not just a silly, ethnocentric superstition policing "typical North American bonds of kinship" (as Villagra [53] might have it)? Could it be that refraining from cannibalism marks, rather, a nearly universal omission that partially constitutes the very idea of kinship—so that advocating the ingestion of "our kind" (whatever boundaries that phrase may encompass) is itself akin to rounding a socionormative square? If one answers these questions affirmatively (as I am inclined to), Villagra's essay appears in the final analysis to want the concept of cannibal-cake without eating the actual practice too—in other words, in reducing to anthropocentrism without anthropophagy, it loses the critical nerve of its own challenge and ends up sounding like yet another (albeit anthropologically sophisticated) rationalization of homo-exclusive carnivory. Hence I am prepared to conclude here that, possibly outside of certain mortuary practices,[10] an anti-speciesist CMV need not tolerate cannibalism—and nor should it embrace the subsistence hunting or humane farming scenarios that usually crop up in the literature on dietary morality.

> *Disgust, then, should be seen from this perspective as a species of wonder, one that is productive at the root of ethics.* (italics added)
>
> Avramescu 2010

Acknowledgments

The author would like to thank Carol Adams and Lori Gruen for accommodating his odd tastes in theory, cultivating intellectual companionship, and making allowances for compositional latitude.

Notes

1   Which is not to say that the tension is totally dissolved: see Russell
    Weiner's stab at a credo of carnivory, "When eaten ... they [who are
    eaten] may occasion a transcendent experience akin to the experience
    of the beautiful, an experience I will call the Delicious. Through
    such an experience the [prey] transcends its abject particularity and
    becomes a vehicle for the universal" (2012); cf. Perry Farrell's lyrical
    confession/conundrum of being caught in food chains, "One must eat
    the other who runs free before him / Put them right into his mouth /
    While fantasizing the beauty of his movements / A sensation not unlike
    slapping yourself in the face" (1990).
2   See especially Warren 2000, Chapter 3.
3   It is worth noting that this episode qualifies her to speak on
    bio-ecological embeddedness from a more even-handed background
    of experience, in contrast to those "holy hunters" (criticized by Marti
    Kheel for faux sanctity) who are wont to extol the eco-immersive
    virtues of adopting predaceous practices without ever balancing their
    outlook by enduring the risks or assuming the vulnerabilities consti-
    tutive of the prey position or role.
4   Plumwood herself nods in this direction when she mounts a critique
    of what she refers to as the "Use Exclusion Assumption" of hardcore
    animal rights theory (2002, 156ff.).
5   E.g. 267 (n. 161): "When, where, and how individuals express their
    ideals are important considerations for a contextual ethical approach."
6   In other words, I do not know that any of the theorists discussed above
    (all of whom are aligned with ecofeminism and/or ethics of care)
    come out in its defense. It should be noted, however, that at least one
    prominent animal ethicist often appears tolerant of or even sanguine
    about non-industrial animal husbandry—but, though Bernard Rollin
    might be counted correctly a contextual moral vegetarian, his discourse
    of animal teleology (rooted in Aristotelian biology) and its flirtations
    with perhaps patriarchal nostalgia for pastoral practices of yore places
    him outside the fold of ecofeminist/care-ethical thought that is the
    prime focus of the present paper.
7   It was loosely based on an earlier novel (Harrison, 1966).
8   For an expose of welfarist farms, see http://www.chooseveg.com/free-
    range.asp
9   As I wrote this chapter, a former officer of the NYPD was on trial
    for conspiracy to cannibalize women. The defense argued that the

1995. Sue Coe's book, Dead Meat, with an essay by Alexander Cockburn, captures the costs of meat eating on animals and slaughterhouse workers.

alleged perpetrator was involved only in fantasy role-playing; to my knowledge, they have not mounted a campaign to exonerate based on religious or otherwise spiritual exemption. The jury returned a conviction, which is being appealed without substantial change in defensive rationale.

10 Actual cannibalism (of the human variety) is hard to investigate and verify—aside from rare anecdotes of psychotic episodes in Western civilization, many stories of indigenous cultures harboring the practice turn out to be apocryphal; the cases with the highest level of corroboration appear to feature the ritual consumption of recently deceased kin (whose death was not the result of cannibalistic intent)—and, depending on the details involved in any given instance, one could mount a defense of such by appeal to Curtin's notion of "authentic presence" to food.

Inter-Animal Moral Conflicts and Moral Repair:

A Contextualized Ecofeminist Approach in Action

Karen S. Emmerman

9

Traditional approaches to conflicts of interest between human and nonhuman animals (inter-animal conflicts of interest, as I call them) generally fixate on offering dilemmas where a human interest is pitted against a nonhuman animal interest in a scenario where only one party can win. These dilemmas typically follow a predictable course where two very clearly defined options based on conventional ways of seeing matters are put forth and we are meant to pick between these two options. We are presented, for example, with an overcrowded lifeboat that for reasons surpassing understanding contains nameless, narrative-less, unknown dogs and humans. Our job is to use moral reasoning to determine which party is tossed overboard, thus ensuring continued life for the other occupants.

Ecofeminist animal theorists have long expressed frustration with this, dominant, approach to moral deliberation.[1] Emphasizing the need for context and narrative, these theorists have noted that inter-animal conflicts take place within social, economic, cultural, and political background conditions. To contemplate a dilemma stripped of the information these background conditions provide is, as Marti Kheel wrote, to engage in a "violence of abstraction" (Kheel 1993, 255). Kheel, like many ecofeminists, suggests that we simply refuse to engage with such questions (259). I agree that talking about cases in these abstract, either/or, ways encourages an impoverished discussion by focusing only on two options and ignoring potentially salient details such as how we found ourselves in the particular dilemma in the first place. Moreover, discussions of inter-animal conflicts of interest tend to set a human interest against a nonhuman animal interest for the purpose of pumping our intuitions about the importance of the former over the latter. But this methodology rests on the dubious notion that approaching cases with such a degree of abstraction is morally unproblematic.[2] An ecofeminist approach resists describing conflicts in these abstracted and unrealistic ways, requiring instead that we tend to all of the relevant features of a case.[3]

Insofar as this is a point about how discussion regarding inter-animal conflicts ought to be undertaken I am in complete agreement. Ecofeminist, contextualized accounts will simply refuse engagement with a question engendered by a false dichotomy and presented without any narrative. Still, I part ways with the tendency to want to focus questions primarily on how we got into the mess we are in, inasmuch as that tendency is unhelpful as practical guidance for moving forward in specific instances of conflict. We are embedded in systems that function largely beyond our immediate control when we are at discrete decision points. These systems are driven by large, often multinational companies, corporate greed, and governments more responsive to powerful lobbyists than a concerned citizenry.[4] Certainly, under these conditions, it is crucial that we ask questions about how we landed in the mess we are in. For a particular mother making a quick decision between a pig heart valve and her daughter's life, asking questions about why the pig's valve is the only available option is part of the broader discussion about what options are available and why. Yet, at the moment of decision, the mother needs to know what to do and how to feel about it.

Ecofeminist animal theory has been the driving force in shifting discussion away from traditional approaches to inter-animal conflicts focused on falsely dualistic dilemmas and truncated narratives. Subsequently, there has been a tendency in the ecofeminist literature to avoid engaging in specific cases. In this paper, I push past this tendency and show how an ecofeminist approach to inter-animal conflicts of interest functions when confronted with particular instances of conflict. We should absolutely refuse engagement with truncated narratives and false lifeboat-style dilemmas. It is also crucial that we attend to what contextual details reveal about exploitation and marginalization and how they factor into creating the dilemmas we face. At the same time, because of the ubiquity of nonhuman animal use and abuse, cases of conflict are the norm rather than the exception. We do not need dramatic lifeboat cases to tell us what we already know, namely that even those of us firmly committed to ending animals' unnecessary suffering are often in situations where all the choices available to us will do violence to animals' interests in some way.[5] An ecofeminist approach to inter-animal conflicts of interest must encourage us to attend to the worldview that created the dilemma. I would add, though, that an ecofeminist theory applied to actual cases helps us see what context can tell us about what to do, how to feel about our moral choices, and what other moral work might remain even if we have done our best in making a choice. This theoretical lens will be useful to activists as well insofar as it may help limit the tendency in the activist community to insist that there are perfectly clean ways out of inter-animal conflicts if only we are committed enough to pursuing them.[6]

Before examining the ecofeminist approach in action, I want to note the critical features of such an approach to inter-animal conflicts of interest. I do not have the room here to provide a robust defense of these features.[7] For the moment, I will simply articulate what they are so that we can think about how such a theory works in practice. As with any theoretical approach, there is no singular set of features that defines an ecofeminist method for thinking about human/animal interactions. The view I offer here is not meant to represent all ecofeminist theories.

Like many other ecofeminist animal theorists, the approach to inter-animal conflicts of interest I endorse is non-hierarchical, pluralist about moral significance, and contextualized, moving between relevant features of the conflict to obtain as full a picture as possible of what is at stake for all parties (Emmerman 2012).[8] This process includes, among other things, detailed descriptions of what is at stake for the humans and the nonhuman animals, situating those details in historical, political, and societal context, and asking hard questions about how we found ourselves in a situation of conflict to begin with. It also involves being aware of interlocking oppressions, how privilege is functioning in our deliberations, and the necessity of careful cross-cultural communication. This expresses what it means for the view to be contextualized. The approach is pluralist in that it recognizes that moral significance arises from a variety of sources (sentience, having a wellbeing of one's own, relationships of love and care, etc.). While early animal theorists working within traditional moral theory (for example, Peter

Singer and Tom Regan) relied solely on reason to account for animals' moral considerability, the ecofeminist approach insists that emotions must have a significant role in determining what matters morally.[9] Finally, the approach is non-hierarchical in that it does not give pride of place to any one kind of interest or creature. My view rejects preconceived moral hierarchies where types of sentient creatures or types of interests are ranked prior to the examination of a given conflict.[10] This means that human lives are not prima facie more valuable than animal lives. Of course, in any given conflict, decisions must be made regarding the various interests and whose will prevail. Rather than endorsing a universally binding hierarchy of interests or life forms, I prefer a methodology that recognizes differences without ranking them in a static, universally binding way. Differences between species and individuals help us understand need and capability but do not provide justification for ranking those species and individuals according to levels of moral significance.[11]

My ecofeminist approach to inter-animal conflicts also highlights the importance of recognizing moral remainders and attending to the work of moral repair when we cause harm.[12] The cost of any human life is that our interests will sometimes conflict with those of someone else. We will often do harm to others' interests. Whether or not we are ultimately justified in our choices, we cannot escape that these choices generate collateral damage for morally significant others. Recognizing that damage, seeing it for what it truly is, and confronting what we might do about it is a crucial part of navigating inter-animal conflicts. As a result of our choices and actions sometimes nonhuman animals will lose, sometimes humans we do not know will lose, and sometimes humans or nonhumans we love and cherish will lose. We have to accept that moral remainders are often a part of moral life even when we do our very best to mitigate all harms. This is as true in the inter-animal realm as it is in the inter-human realm.[13] That we made the best choice of the options available to us most often will not mean that we can say "done, clean, finished" once we have chosen the best available alternative. Moral work will likely remain. As Margaret Urban Walker put it,

> But if moral life is seen as a tissue of moral understandings which configure, respond to, and reconfigure relations as they go, we should anticipate residues and carry-overs as the rule rather than the exception: one's choice will often be a selection of one among various imperfect responses, a response to some among various claims which can't all be fulfilled. So there will just as often be unfinished business and ongoing business, compensations and reparations, postponements and returns. (Walker 1995, 145)

A pluralist, non-hierarchical, and contextualized approach to inter-animal conflicts of interest helps us recognize the complexity and plurality of the interests at stake *on all sides* of a conflict and it forces us to take a more honest look at our dealings with others. Moral life is in large part about recognizing remainders as the norm, rather than the exception.[14]

Any approach to inter-animal conflicts of interest must take these realities seriously and address the issue of remainders head-on. A view that does not is out of touch with reality.[15] As Virginia Held argues, theories must be tested against actual, lived experience (Held 1995). Lived experience tells us that, in many cases, we do not simply maximize the good, respect rational agency, or show loving attention and move on worry-free even if moral theory tells us we could. We know moral remainders are a part of moral life because we experience them. An adequate theory will have something to say about this experience beyond thinking of it as mere squeamishness or sentimentality.

With these features of an ecofeminist approach in mind, we can now turn to thinking about what such an approach might have to say about particular cases of conflict. There is a methodological tension here, to be sure. A central feature of an ecofeminist account is that it treats interests as situated in social, political, cultural, economic, and relational context. This means that too much discussion of hypothetical and inevitably truncated cases is antithetical to the approach. I have opted to cope with this delicate balance by focusing on an autobiographical example where I have as much under-standing of the contextual complexities as possible. I do not need to operate with a truncated narrative. I will discuss the dilemma of feeding my newborn son. Here is the case:

In 2006, I gave birth to my son prematurely at 33 weeks gestation with no previous warning signs. I had been hoping to breastfeed my child, but like some mothers of premature infants I was unable to do so.[16] I was provided with a breast pump by the hospital and coached by lactation consultants but, after two weeks of pumping every three hours day and night with no results, the consultants suggested that I pursue other options. During those two weeks my son had been receiving formula. I am a vegan and was committed to raising a vegan child so I insisted that the neo-natal intensive care unit (NICU) provide my son with soy formula. In 2006 it was impossible to obtain vitamin D3 from non-animal sources. No infant formula is made without including D3 in the mix. I was able to find a soy formula that sourced the D3 from sheep lanolin and opted for that as the best I could do.

I am a person who strives as much as possible to live a life free from causing violence to other living beings. Sheep raised for wool receive terrible treatment and suffer a great deal over the course of their lives.[17] The situation caused me considerable distress. What does an ecofeminist approach have to say about this?

Because the approach is non-hierarchical, I cannot reduce the conflict to my son having more moral significance than the sheep used to make his formula. My son is more significant *to me* than the sheep because he is my son and I am his mother and this clearly influences my choices. He is not more morally significant than the sheep, however, simply because he is human. So, it won't do to say that in a conflict between feeding a human child and harming a sheep, the child ought to prevail because he is morally more significant than the sheep.

Still, I am certain it seems painfully obvious to everyone that opting not to feed him anything at all as a form of protest against the lanolin-derived D3, thus leaving my son to die of starvation, is not an option. This is because, like the sheep, he is also a living, breathing, sentient being in relationship with others (to the extent possible at birth). Also, he is my son, which means I have the responsibility to see to his needs and ensure his survival wherever appropriate.[18] The fact that he is my son also means that he is not a stranger with whom I have no connection. He is a beloved, much-anticipated member of my family.

Ecofeminist animal theorists would want to ask a few questions here. We might ask, for example, why we have not developed a vegan source of D3 and is that because we take for granted that using sheep in these ways is morally unproblematic?[19] Was there an environmental reason, caused by morally dubious behavior of another sort, for why I gave birth to a premature infant in the first place that should be addressed? These questions are reasonable and important. They also would not likely be brought to light in a traditional approach to inter-animal conflicts that focused entirely on simply categorizing the interests at stake or on defending a moral hierarchy of life forms. These questions need to be explored and addressed, but their answers were not going to help me, given the pressing nature of the situation.

At the time, I opted to give my son the soy formula with lanolin-sourced D3 while recognizing that in doing so I generated moral remainders with respect to the sheep. While that is what I *did* do, I have often wondered if that is what I should have done. A guiding principle of my approach is that we ought not to harm sentient life unless by doing so we seek to benefit them or defend ourselves.[20] Therefore, using the lanolin-derived D3 formula is deeply problematic. Now, this approach is meant to strike a balance between principles and particulars so it also requires that I note the following things: (1) I could not breastfeed despite trying my best; (2) my son needed to eat something or he would die; (3) I am embedded in a context that for scientific, cultural, and speciesist reasons has not generated a completely vegan infant formula; (4) I was exhausted from his birth and medically compromised myself, so not at the pinnacle of my moral powers; and (5) I was making these decisions in the context of an intensive care scenario which presented both structural and emotional complications. These are reasons I do not retrospectively cast tremendous blame on myself for the decision I made at the time.

Still, this does not mean that I could justify going to any length to provide him with nutrition. If the doctors had suggested that the only way to keep him alive was to kill the baby in the next isolette, I could not have accepted that option. There are limits to what we can do for our loved ones, to be sure. But this points to the interesting way in which distance factored into my judgment at the time. Had the doctors hauled a sheep into the room and said, "Now, we're going to cut off this sheep's skin without anaesthesia in order to best obtain the lanolin from her wool so that we can feed your son," I would have demanded alternative solutions. Respect for the sheep's moral significance and empathy for the sheep's desire not to suffer would have required that I had done so, just as respect and empathy for the stranger's baby in the next isolette requires that I demand alternative solutions in that hypothetical case.

The most obvious alternative would have been to rely on human breast milk donation. Unlike in the past, when wet nurses were often coerced or forced into providing milk for wealthier families (or slave owners in the case of antebellum "mammies") at the expense of their own infants' health, contemporary breast milk donation is entirely voluntary. Women with surplus milk donate in order to help those who need their milk, thus alleviating concerns about exploitation. The donated breast milk alternative is obvious now, but the contextualized nature of my approach reminds us to evaluate moral dilemmas in their historical context. In 2006, I was unaware of any discussion of breast milk donation either in the broader community where I live or in the hospital where my son was treated. Indeed, my son's NICU did not start using human breast milk banking until October 2012.[21] It is difficult to identify precisely what prevented the NICU from using human milk banking for their premature infants. It seems to have been a combination of concerns regarding cost, a lack of in-state milk banks, and that breastfeeding was not in itself something the hospital staff deemed particularly important.[22] Had the hospital offered donated milk through a formal breast milk banking system I imagine I would have accepted their offer. The assurance that the donors had been screened and the milk pasteurized would have reassured me of its safety.[23] Even so, my son's relative healthiness for a premature infant would have disqualified him from receiving milk from a donor bank. Human milk from donor banks is expensive and offered to those babies most at risk for particular medical complications.[24]

1996. *Beyond Animal Rights: A Feminist Caring Ethic for the Treatment of Animals*, edited by Josephine Donovan and Carol J. Adams, published.

What if someone had offered milk through an informal sharing system where women with surplus breast milk donate to mothers and infants in need? In 2013 this sort of casual sharing of breast milk is becoming more and more common. There are social networking pages where mothers in the community post that they have a surplus of milk and those needing milk can arrange to acquire it.[25] In 2006, this kind of informal system for donating breast milk was not a part of the parenting culture where I live. Thus, it never occurred to me to even ask about breast milk donation at the time. The forces working against casual breast milk donation were not merely historical, however. There were also structural and emotional impediments.

On a structural level, NICUs differ from one another in how they respond to families wanting to use unscreened, donated milk. I suspect that, in 2006, any such request would have been met with significant resistance from my son's medical team.[26] Still, in my case, there would have been limitations beyond these structural issues. Had the hospital permitted me to use milk obtained through casual donation, I cannot imagine that I would have taken advantage of the opportunity. Mothering an infant in the NICU puts a person in a very fearful place. Thoughts about death and the fragility of life are inescapable in that setting even if the prognosis for one's child is good. Given that I already feared for my son's life, I would not have accepted the potential risks involved in exposing his premature immune system to unscreened milk. The thought that I might make a choice that resulted in his death would have been too much for me to bear in that context.

The contextualized nature of my approach reveals the ways my options were limited by the conditions in which I found myself. Situated as I was, in 2006, soy formula was truly the only reasonable alternative for feeding my son in a way that came as close to conforming to my vegan values as possible.[27] We may be able to say that I truly made the best moral decision I could at the time. At the same time, this approach also sheds light on the moral remainders created by my decision and the necessity of looking to the work of moral repair.[28] It makes clear that I must circle back to the questions raised earlier about why we live in a world that treats nonhuman animals as resources, why no vegan infant formulas exist, and what, if any, environmental factors contributed to my son's premature birth in the first place. I must also return to thinking about the sheep. In this situation it is impossible to know which or how many sheep were affected. It is thus untenable to suggest that I make some sort of restitution to that (or those) sheep. Depending on my situation and resources, it may mean that I donate money to an organization working to improve the conditions of farmed sheep or to a sanctuary working with rescued sheep. There are a variety of options for what would be appropriate depending on the particulars of the situation.

Built into my thinking, however, must be the realization that I can never make restitution to the particular animals harmed in this situation and that should give me pause. This is not to suggest that I ought to be consumed with guilt and suffering as a result, but that I should approach the situation with some gravitas and recognition that my son's life was maintained through the sacrifice of others' wellbeing.[29] This acknowledgment and appreciation will hopefully fuel the fire of my resolve to both change the mechanisms in place that forced me into such a dilemma in the first place and support those seeking to improve animals' lives.

Ecofeminist animal theories have historically insisted on the importance of context in part to make clear that moral conflicts do not arise in a vacuum. There are reasons vegan mothers who cannot breastfeed are in a position where they have to choose between their commitment to living without causing violence to nonhuman animals and the ability to feed their children. These reasons can be explained by pointing to, among other things, the fact that we live in a society that systematically exploits nonhuman animals for human purposes. Therefore, for very good reasons, ecofeminist animal theorists have advocated against allowing false, dualistic dilemmas to drive our thinking about inter-animal moral conflicts. I hope to have shown here, though, that an ecofeminist approach to inter-animal moral conflicts can move beyond both conversations about how we got into these conflicts in the first place and the rejection of truncated narratives. Attending to context can reveal the underlying political, social, and cultural forces that limit our choices, but doing so can also help us identify moral remainders and undertake the work of moral repair. The inter-animal conflicts of interest we face may often not be resolvable in ways that are satisfying to our sense of justice or our feelings of compassion and care. My approach hopefully moves us a step forward in offering ways of thinking about these conflicts that captures this insight by highlighting the important role moral remainders and moral repair play in our moral lives.

Afterword

As of this writing (January 2013), there are now vegan D3 supplements available for purchase.[30] Currently, there are no vegan infant formulas available in the United States or the United Kingdom as all brands continue to rely upon lanolin-sourced D3.[31] We can hope that the companies making formula for infants will make changes as the availability of vegan D3 expands. Applying pressure on formula companies to make this change would fit into my conception of the work of moral repair for those of us who have had babies whose lives were made possible by the exploitation of other creatures.

Notes

1   I use the language of "ecofeminist animal theory" because not all ecofeminist theory is inclusive of animals' concerns. Greta Gaard uses "vegetarian ecofeminism" to clearly mark ecofeminist theory that takes the oppression of animals as integral to ecofeminist work (Gaard 2002). Throughout this piece I refer to an "ecofeminist approach" as shorthand for an ecofeminist animal theorist approach, since the latter is cumbersome. It should be clear to readers that I am offering an ecofeminist analysis that is deeply inclusive of animals' interests.

2   Lori Gruen discusses the problems with undertaking ethical inquiry through lifeboat cases (Gruen 1991, 352). Cf. Francione 2004, 133; Mellon 1989, Chapter 3.

3   Cf. Carol Gilligan's discussion of the Heinz case (Gilligan 1982, 25–31). Presented with a situation where Heinz either had to steal a drug he could not afford or watch his wife die without it, many women and girls wanted more information about why Heinz could not work in exchange for the drug, ask the pharmacist for a loan, and so forth. Presented with only two options (steal the drug or let one's wife die), these respondents demanded more information.

4   To some extent, this is what Rosalind Hursthouse is getting at when she argues that Peter Singer and Tom Regan's claim that animal research is wrong is not action-guiding for most of us in certain circumstances (Hursthouse 2006). Aside from actual vivisectors and lab technicians, it is unclear what it would mean for us to opt out of animal research. Of course, we can eschew products tested on animals and that sort of thing, but to the extent that the medical industrial complex is so much more powerful than we are and limits our choices in significant ways, opting out of certain aspects of animal research is sometimes impossible. A premature infant who was intubated at any hospital until very recently benefitted from the fact that endotracheal intubation has been practiced on small animals, like ferrets, in pediatric training hospitals for years. That research is, thankfully, being changed to non-animal models. But for parents in the neo-natal intensive care unit it is difficult to know what to do with the general guidance that

we should oppose animal research when in the actual moment of decision. Clearly, options for activism and awareness-raising abound, as Hursthouse points out, but the issue remains that we often cannot opt out of certain practices in the immediate situations we are in.

5   Cf. Curtin 1991.

6   Here I am reminded of the advice I once received on a vegan listserv. I had asked if anyone knew of a vegan source of vitamin D3 and the response I got was that there were not any vegan supplements but I should go outside in the full sun without applying sunscreen. The person also suggested that people are far too serious about sunscreen application. This struck me as a willful denial of the reality that, in the absence of vegan D3 supplements, there may be a genuine moral dilemma at play.

7   See Emmerman 2012 for a full discussion and defense of the features of the ecofeminist approach I mention here.

8   There are many examples of pluralist, contextualized, and non-hierarchical approaches to animal ethics. See Curtin 1991; Donovan and Adams 2007; Donovan 1990; Gaard 1993, 2001, 2002; Gruen 1993, 2004, 2011; Kheel 1985, 1993; Luke 1995, 2007; Slicer 1991, among others.

9   See Singer 1990 and 1994 and Regan 1983 for their reason-only approaches to animal ethics. When I refer to traditional animal theory I reference these kinds of views. For excellent discussions of the importance of emotion to moral deliberation see Donovan 1990; Gaard 2002; Gilligan 1982; Gruen 1991, 1993, 2004; Held 1995; Luke 1995, 2007; Kheel 1985, 1993; Slicer 1991; and Walker 1989, 1995.

10   For two examples of theories that rely on such a hierarchy, see Varner 2012 and Warren 1997.

11   In future work I will think more about what should be said about non-sentient living things. I am keenly aware of the dangers of replacing one form of hierarchy with another and would very much like to avoid doing so (Kheel nicely articulates these dangers in Kheel 1985). Still, it is not yet clear to me exactly how to talk about non-sentient living things.

12   I borrow the language of "moral repair" from Margaret Urban Walker (Walker 2006).

13   Moral hierarchies that rank sentient creatures according to levels of moral significance are problematic in part because they have taught us not to see the remainders where our interactions with animals are concerned. If we think of a cat as less morally significant than a human, then we feel justified in allowing a human's interest to trump

1998. "Ecofeminism: A Practical Environmental Philosophy for the 21st Century" Conference held in Missoula, Montana.

the cat's interest. That feeling of being justified in turn discourages us from seeing that there may be more work to do with respect to the cat when her interests are thwarted in cases of conflict.

14 I am not suggesting that there are *always* moral remainders. Certainly, if I make a promise to a friend and I keep that promise, then it is likely that I will not have generated any moral remainders in that process. In inter-animal conflicts, however, I suspect that moral remainders will more likely be the norm rather than the exception. We are embedded in exploitative systems over which we have very limited (or no) direct control. As a result, there may not be good ways to extricate ourselves from inter-animal conflicts without causing significant harms. Moreover, given that attending to animals' interests in a robust way is still on the fringes of common morality, it will be difficult to attend to animals without generating some problems in the inter-human realm, in particular with people we love or other community members. One need only read blogs written by vegans at Thanksgiving to see that this is the case.

15 Sharon Bishop has a thorough discussion of how principle-based theories handle (or fail to handle) feelings of remorse and guilt in Bishop 1987. I am trying to move away from the sort of view Bishop criticizes to one that can more easily recognize remorse and the appropriateness of acts of forgiveness or repair.

16 My situation was complicated by physiological issues unrelated to my son's prematurity. Many mothers of premature infants are able to breastfeed successfully with adequate support.

17 Gene Baur says, "Even in the production of wool, cruelty is a feature. To reduce problems with flies that infest the folds in the skin of Merino sheep ..., producers practice 'mulesing.' Strips of flesh are literally cut off the backs of the animals' legs and hind region to create smooth skin without anesthesia or pain relievers. Sheep also commonly have their tails cut off to control fly problems" (Baur 2008, 79). This treatment is not limited to Merino sheep (see Martindale 2010). There is also a connection between sheep used for wool and sheep used for consumption. When sheep "age out" of producing wool, they are slaughtered for human consumption and the lambs born of sheep in the wool industry are shipped and slaughtered for human consumption, sometimes being shipped alive for very long distances. Because of declines in demand for wool, sheep are being genetically modified for dual-purpose use so that they are both good wool producers and good sources of "meat" (Jones 2004).

18 I say "wherever appropriate" because there may be cases, as if he

is terminally ill, where ensuring his survival is either impossible or undesirable.

19   To be fair, there is a non-animal source of D3. It is the sun, but clearly sunshine is not an option for most infants and is certainly not an option for neonates living in isolettes in intensive care units. Recently, a vegan source of D3 has been formulated. I will say more about this in the afterward at the end of the paper.

20   Josephine Donovan and Carol Adams have offered the following core principles for animal ethics: "It is wrong to harm sentient creatures unless overriding good will result for them. It is wrong to kill such animals unless in immediate self-defense or in defense of those for whom one is personally responsible. Moreover, humans have a moral obligation to care for those animals who, for whatever reason, are unable to adequately care for themselves, in accordance with their needs and wishes, as best as the caregivers can ascertain them and within the limits of the caregivers' own capacities. Finally, people have a moral duty to oppose and expose those who are contributing to animal abuse (Donovan and Adams 2007, 4). The view I offer here is informed by Donovan and Adams, though I might alter the details of their principles somewhat.

21   Ginna Wall, RN, MN, IBCLC, e-mail to author, 14 December 2012.

22   Apparently, it was not until the 2011 Surgeon General's report on breastfeeding was released that there was an "official" recommendation to use donor milk (Wall, e-mail to the author, 20 December 2012). Clearly, this was not an impediment for some hospitals, as the Mothers' Milk Bank of California has been providing NICUs with donated milk since 1974 (http://www.sanjosemilkbank.com/about.htm [accessed 12 Jan. 2013]). The La Leche League has a brief history of donor milk banking on their website that might be of interest to some readers (https://www.llli.org/llleaderweb/lv/lvaprmay00p19.html [accessed 18 December 2012]).

23   Of course, whether a vegan should accept breast milk from a non-vegan donor is an open question. This issue has been debated on some vegan parenting sites (see http://www.mothering.com/community/t/671974/vegan-infant-formula [accessed 19 December 2012]). This issue highlights how difficult it can be to fully extricate oneself from inter-animal conflicts.

24   Human milk obtained through the milk banking system costs $3.00 per ounce compared to $0.15 per ounce for formula. Hence, donor milk is "used like an expensive medicine" (Wall, e-mail to author, 17 December 2012).

1998. Randall Lockwood and Frank R. Ascione's *Cruelty to Animals and Interpersonal Violence: Readings in Research and Application* published.

1998. Jennifer Abbott's *A Cow at My Table* premieres.

25   See, for example, Human Milk for Human Babies at www.hm4hb.net
     My thanks to Jennifer Mendelson for bringing me up to date on current
     trends in casual breast milk donation.

26   The current policy is to allow parents to feed their infants milk
     obtained through unscreened, casual donation after they have been
     informed of the risks by the medical team (Wall, e-mail to author, 20
     December 2012). Even though casual donation is permitted, some
     physicians strongly discourage parents from using unscreened human
     milk donations.

27   I do not doubt that vegan mothers more informed about breast milk
     donation and less frightened of immunity concerns may have pursued
     other options. I do not mean to suggest here that no other option
     would have been reasonable, only that the one I pursued seems under-
     standable under the circumstances.

28   Though I cannot go more deeply into this here, the work of moral
     repair may amount to different things depending on the kinds of
     dilemmas we face. Some dilemmas come about because of deeply
     entrenched social and cultural structures, while others cannot be
     resolved by making changes in those structures. My analysis of
     my own situation is that it was a bit of both. There were structural
     limitations around the acceptability and availability of breast milk
     donation in a NICU setting as well as limitations created by the
     kind of fear and doubt many people suffer when in complex medical
     situations.

29   This kind of recognition can be hard to articulate. Deane Curtin's
     discussion of contextualized moral vegetarianism helps shed light on
     what I am getting at here. In discussing the fact that some cultures, due
     to extreme weather conditions, cannot follow a vegetarian lifestyle,
     he notes that many have rituals in place that involve paying respect to
     the animals harmed or killed. He says that "In some cultures, violence
     against nonhuman life is ritualized in such a way that one is present
     to the reality of one's food" (Curtin 1991, 70). These rituals show
     mindfulness about the consequences for others of the choices we
     sometimes have to make. It is possible that, in some cases, the rituals
     express gratitude to the gods or the universe, but do not reflect the
     idea that anything wrong has happened. I have both senses of ritual in
     mind here: those that reflect a regretful attitude and those that express
     gratitude. Mindfulness or "being present to the reality" is an important
     part of having an appropriate mindset regarding moral remainders. Our
     culture's current practices reflect neither respect nor gratitude towards
     the animals sacrificed in service of human ends.

30 There is some disagreement about whether all of these supplements are strictly vegan, but at least one brand appears to pass muster for the Vegan Society. See the discussion on Jack Norris's website regarding vegan D3 supplements: http://jacknorrisrd.com/?p=2081 (accessed 28 December 2012). Norris is a vegan nutritionist and the founder of Vegan Outreach.

31 My thanks to lauren Ornelas of Food Empowerment Project for helping me with this research.

The Wonderful, Horrible Life of Michael Vick

Claire Jean Kim

10

It was April 25, 2007, when the Surry County Sheriff's Department executed a search warrant at 1915 Moonlight Road near Smithfield, Virginia, a 15-acre property owned by NFL star Michael Vick.[1] Investigators found more than 50 dogs, many of whom were scarred or wounded, and the standard paraphernalia of dogfighting including a breeding stand ("rape rack"), treadmill, breaksticks, and injectable steroids (Strouse 2009).

Thus began Michael Vick's spectacular fall from grace. Vick had gone from humble beginnings in a Virginia public housing project to NFL superstardom. He was the first African American quarterback to be selected first in the NFL draft (by the Atlanta Falcons in 2001) and was one of the most talented and promising players in the league. Just before his fall, he was flying high with a ten-year, 130 million dollar contract with the Falcons and endorsement contracts with Nike, Coca Cola, Reebok, and many others (Laucella 2010). Then one of Vick's cousins was arrested for marijuana possession and gave 1915 Moonlight Road as his address, leading police right to Vick's illegal dogfighting operation. Prosecutors charged Vick and three of his friends with several federal felony counts relating to the operation of an interstate dogfighting and gambling ring known as Bad Newz Kennels. For months, Vick lied to investigators, claiming that his friends had engaged in dogfighting on

2000. Eastern Shore Sanctuary (now VINE) founded.

his property without his knowledge. As the evidence against him mounted, he was suspended from the NFL, lost endorsement contracts, and filed for bankruptcy. Finally, after his three friends stated they were ready to give evidence against him, Vick pled guilty and was sentenced to 23 months in prison. But the rags to riches to rags story had one more twist. After Vick was released in 2009, after serving 18 months, he worked to redeem himself in the eyes of the NFL management, corporate sponsors, and the public. Working with several mentors and a team of more than seven public relations advisors, Vick resurrected his image (by acting as a spokesman against dogfighting for the Humane Society of the United States) and his career (he joined the Philadelphia Eagles in 2009 and once again signed a contract with Nike). Vick also won the Ed Block Award for courage and sportsmanship and the Associated Press NFL Comeback Player of the Year Award.

The American Dream is the simple idea that if you work hard, you can make it, with making it defined quite specifically as the attainment of middle-class status or more and symbolized by ownership of a single family home. The adjective "American" here is a qualifier meant to express not humility—the recognition that people elsewhere might dream other dreams—but rather exceptionalism. This dream is only possible here, has been crafted to perfection here. Vick's story of rags to riches to rags to redemption has been read by many observers and by Vick himself as a confirmation of the American Dream. For some time, professional sports has been touted as a site of successful racial integration in the US, a theater of black accomplishment and postraciality, and Vick's story has been read through this lens. First he triumphed over the projects, then, humbled by his own inner demons, he fought his way back to NFL stardom and wealth. Only in America could Michael Vick have written his own story not once but twice.

I would like to challenge this reading of the Michael Vick story and ask what it willfully, perversely forgets. The Michael Vick story, if we attend to it closely, is not one about self-realization in the context of freedom and opportunity but rather one about how taxonomies of power such as race, species, and gender constrain and indeed incarcerate the bodies they produce. The Michael Vick story does not validate the American Dream but rather exposes it as a fable. In the 1800s, the promise of America was used to manage the contradictions of racial and class inequality in nascent capitalism in order to produce quiescent workers and citizens. Now, in this neoliberal age, as these contradictions sharpen, the fable of the American Dream is more socially necessary and more empirically untrue than ever.

Michael Vick's arrest in 2007 ignited a firestorm of media commentary and public debate on television, on the radio, and in cyberspace. From the start, Vick's defenders argued that race was driving the prosecution. Animal advocates, many of whom are white, forcefully denied this, insisting that the case had "nothing to do with race" and that Vick was under fire only because of his cruel treatment of dogs. One caller on NPR's *Talk of the Nation* said: "I don't care if Michael Vick was black, white, green, purple. To me, this is not a story about color," and PETA's blog on the Vick case stated unequivocally: "This is not a race issue. We don't care if he's orange. This is not a race issue. White people who fight dogs need to fry. This is not a race issue."[2]

Given the historical period we inhabit right now, it is not surprising that this became the central issue—was the Vick case about race or not? This period, which Eduardo Bonilla-Silva (2009) calls the period of "color-blind racism" and Patricia Hill Collins (2005) has called the period of "new racism," is marked by the contradiction of official race-neutrality and discourses of colorblind transcendence on the one hand, and ongoing racialization on the other. A period when race is everywhere and everywhere denied. A period when the nation's first African American president is denounced as a cocaine dealer, an affirmative action baby, a foreign-born Muslim terrorist, the n-word, and a chimpanzee and at the same time held up as a symbol of our postraciality. In this period, racial justice advocates have become full-time sleuths, ferreting out and exposing racial discrimination as it disguises itself in ever subtler forms, while many whites object that this is not uncovering racism as much as it is "injecting race" into situations or "playing the race card."

Race is everywhere in the Vick story. We live in a society where racial meanings are ubiquitous and pervasive, saturating our thoughts, our speech, our actions. Animal advocates claimed that the Vick case had "nothing to do with race," that they could bracket out race and assert a universalist narrative about cruelty without racial implications. But this is a claim to a kind of racial innocence that is impossible. There is no race free space. Vick's critics said they didn't care if he was green, purple, or orange, but no one in this country has ever been enslaved, auctioned off, or lynched for being green, purple, or orange. Race set the context for the Vick story, shaping poor urban neighborhoods like the one where Vick grew up, where the (mis) opportunity structure makes social mobility hard and the hostile criminal justice system makes going to prison easy—with sports an avenue of escape only for the lucky few. In Ruth Gilmore's (2007) unforgettable words, race still means differential vulnerability to premature death. Driving while black, walking while black, breathing while black remains an offense that is sometimes punishable by death, and not only by the police but sometimes by overzealous (vigilante) citizens as well. We see race in the broader political backdrop, too, where there is an intensified effort to pass voter ID laws and

2000. Clifton Flynn's "'Woman's Best Friend: Pet Abuse and the Role of Companion Animals in the Lives of Battered Women" published in *Violence Against Women.*

otherwise curtail voting rights (or as Justice Scalia calls them, "racial entitle-ments") in the wake of *Shelby County v. Holder* (2013).

I want to focus on the racial meanings which course through the Vick story, and specifically on the way in which racial meanings draw upon and are entangled with species meanings here. One reason the Vick case has been so explosive is that it raises questions about the human/animal boundary, about the intersection of the taxonomies of race and species, and specifi-cally the relationship between black maleness and animality. What did it mean that Vick's critics often discussed his dogs in human terms—as having been executed, as experiencing redemption? What did it mean that some of Vick's critics called for him to be neutered? What did it mean that satirists often reversed the positions of Vick and his dogs, suggesting that Vick was more of an animal than they were? Consider some of the key dualisms which underwrite Western culture: master/slave, male/female, human/animal, white/nonwhite, reason/nature, culture/nature, civilized/savage, mind/body, subject/object. Critical race theorists, ecofeminists, and many others have persuasively argued that these differences are not produced in isolation but are co-constituted or produced as effects of power in a profoundly interde-pendent way (McClintock 1995; Plumwood 1993). The black man is made through synergistic taxonomies of power—not just as not white but as savage, nature, other, body, object, alien, slave, and animal.

From the Middle Ages onward, Europeans viewed black people as the lowest type of humans, ranked just above apes on the Great Chain of Being. The boundary between the two was a matter of great interest and fierce debate, and the black man was variously characterized over the centuries as an ape, as ape-like, or as closest to the ape. Winthrop Jordan (1968) recounts that British explorers in the early 1500s first encountered Africans and apes at the same time and began describing Africans as apes who were tailless and walked upright. In *Notes on the State of Virginia* (1794), Thomas Jefferson wrote that Negro men preferred white women to the same measure that the "oran–ootans" (he was referring to chimpanzees) preferred Negro women, the males of each species reaching up the chain of being in their sexual desires. Decades later, as standard-bearers for the "American School" of anthropology, Josiah Nott and George Gliddon explicitly located Africans midway between Europeans and apes in *Types of Mankind* (1855). Along with Samuel George Morton, they used data from craniometric investiga-tions to not only rank races (whites on top, Indians in the middle, blacks on the bottom) but to argue that they were different species altogether.

Scroll forward a century and a half to Barack Obama. Are we now in a postcivil rights postracial era where these ideas about black masculinity no longer pertain? I recently argued that we saw something new in the

2008 presidential campaign—blackness being recuperated into foreignness, as evidenced by the relentless constructions of Obama as a foreign-born Muslim terrorist (Kim 2011). But there has been plenty of good old-fashioned animalization of Obama as well. In 2012, federal judge Richard Cebull of the US District Court in Montana circulated an email entitled "A Mother's Love," wherein the little boy asks "Mommy, how come I'm black and you're white?" and the mother responds, "Don't even go there, Barack! From what I can remember about that party, you're lucky you don't bark!" During the 2008 campaign, T-shirts and toys featuring Obama as Curious George were on sale online, as were Obama sock puppet monkeys. In February 2009, referencing the fight over the president's stimulus bill and the police shooting of a pet chimpanzee in Connecticut, a *New York Post* cartoon by Sean Delonas shows two police officers standing with a smoking gun over a dead chimp, saying "They'll have to find someone else to write the next stimulus bill." In April 2011, Orange County GOP Central Committee member Marilyn Davenport emailed to friends a picture of a "family" portrait of chimpanzees with Obama's face photoshopped onto the baby chimpanzee's body.[3]

In the US, sports, like politics, has been a high-voltage site for the production and circulation of racial meanings. The NBA, NFL, Nike, Gatorade and other corporate interests have sought to package and sell Michael Jordan, Michael Vick, and other superstars as exemplifying the American Dream and the transcendence of race through athletic achievement. But there is now a rich body of scholarship laying out how professional sports, and the NBA and NFL in particular, reinscribe patterns of white supremacy (Coakley 1998; Leonard and King 2011; Leonard 2010; Andrews 2001). Beyond stacking (the assigning of team positions based on race), the preponderance of white coaches, managers, and owners, and racism exhibited by fans at the games, the discourse we use to talk about and think of black athletes bears the imprint of centuries of racial imaginings. As John Hoberman (1997) argues, we talk about natural black athletic superiority even though this insinuates natural black intellectual inferiority. We are more likely to refer to white ball players as smart, savvy, having good judgment or making good decisions. The same goes for honorary white players. In "The Great Yellow Hope," I argue that Jeremy Lin (formerly with the New York Knicks, now with the Houston Rockets), who is Asian American, is heavily praised for his intelligence, diligence, organization, and preparation—all model minority traits.[4] Black male athletes, on the other hand, are continually produced as either good blacks or bad, either transcendent Michael Jordans or incorrigible Latrell Sprewells (Ferber 2007). Both can be depicted as beasts on the verge of being out of control. Consider the 2003 Nike television ad which juxtaposed gritty urban scenes of basketball faceoffs with a pitbull facing

2001. Karen Davis's More Than a Meal: The Turkey in History, Myth, Ritual, and Reality published, showing how both wild and domestic turkeys are subject to human sexual violence.

off against a Rottweiler in a fight. These imaginings of black male athletes at once express collective anxieties about the threat of racial "progress" on various fronts and satisfy the neoliberal imperative of selling more stuff by showcasing the black body as a site of both terror and desire. According to Thomas Oates (2007), we can look at the NFL draft itself—with its preliminary convention of having young prospects strip to their shorts, get weighed, and parade in front of scouts and media—as a disciplinary ritual which feminizes and animalizes these young men, rendering their blackness safe for white consumption.

David Leonard writes: "Black bodies, even those living the 'American Dream,' functioning as million-dollar commodities, are contained and imagined as dangerous, menacing, abject, and criminal" (2010, 259). I would add, as animal. Many cartoonists approached the Michael Vick story through the theme of human/animal reversal. In one cartoon, Michael Vick is pictured next to a fighting dog and the caption reads: "Pop Quiz: Find the True Animal." In another, a dog sitting curbside gives Vick the finger as he is driven away in a van labelled "Animal Control." Another, referencing the fact that Vick electrocuted some of the dogs he found lacking, shows Vick sitting in what looks like an electric chair in a tub of water while a dog holding electrical wires approaches. What are we to make of this image? Is the artist ignorant of the impact of race upon every aspect of the US criminal justice system, from profiling to arrest to prosecution to conviction to sentencing to postsentencing consideration? Does s/he know nothing about the intense public debate raging about how the death penalty is enacted in a racially discriminatory way? What about journalist Tucker Carlson who said on Fox News that Vick should have been executed for being cruel to dogs? What are we to make of these statements about Vick?

Many of Vick's defenders voiced the opinion that whites feel a special glee in bringing down famous and wealthy black men, especially athletes. Mike Tyson, OJ, Kobe Bryant, Barry Bonds, and now Michael Vick. Pointing to a tortured history of police violence against black Americans and a discriminatory criminal justice system, they argued there was a rush to judgment and a harshness of condemnation in the Vick case. This is why some black people talked about the need to close ranks around Vick. R. L. White, head of the Atlanta NAACP, expressed ongoing support for Vick, as did Southern Christian Leadership Conference president Charles Steele, who was quoted as saying: "We need to support him no matter what the evidence reveals." Rev. Joseph Lowery, a veteran of the Montgomery Bus Boycott and other civil rights protests, said from the pulpit: "Michael Vick is my son. I've never met him. But he is my son." Kwame Abernathy, Ralph Abernathy's son, called what happened to Vick an "electronic lynching" (Thompson n.d.).

Polls showed that black Americans were more likely than whites to think Vick's punishment was too harsh and to support his reinstatement to the NFL after his release from prison, although the former were far from unanimous on these issues.

We can only read Vick's story as the American Dream if we willfully, perversely forget race. But race is not the only taxonomy of power that figures large in Vick's story. There is also species—and here I mean the categorical line-drawing we do between humans and all other animals, as well as the way we characterize species natures and breed natures. Like race, species is a constructed system of power built upon the politicization of physical difference. Like race, it is a classificatory exercise with political ends. Species meanings have been an instrument for arrogating to ourselves as humans mind, reason, subjecthood, moral consideration, and rights—and denying these to all other sentient life. Virginia Anderson (2006) points out that seventeenth-century Algonquian-speaking Indian tribes in the Chesapeake area had no single word for "animal," suggesting that they perceived a world of infinitely varied creatures. But the English colonists at Jamestown used the rubric term "animal" to refer to all those creatures whom they saw as less than human. Nonhuman animals do not figure at all in the American Dream, of course, except perhaps as pet commodities signifying middle-classness, along with the house and the picket fence (notably, the first thing Michael Vick wanted after his redemption was a family dog). The American Dream is anthropocentric in two senses: it attends only to the wellbeing, needs, and desires of humans; and it imagines the thriving of humans to be independent from the wellbeing of other species and nature— that is, it ignores the biological, ecological context in which human life and labor takes place.

Many of Vick's defenders resisted the animal question as distracting from the race question, as though it were a competitive, zero-sum situation—race or species, black people or animals. Dogfighting is wrong, they intoned, but in the end, they're only animals. Consider this NPR exchange between host Allison Keyes and scholar Michael Eric Dyson, both of whom are African American:

> Dyson:
> Lassie stayed on the air for 15 years, Nat King Cole couldn't stay on his show for six months. Dogs and animals have been treated—relatively speaking—with greater respect and regard … than African American people. When you look at Hurricane Katrina, they have a famous picture of a bus full of dogs and animals being treated to first-class citizenship rights in America while black people were drowning … this is not to disrespect

the needs of other sentient animals who would coexist with us on the human space called earth.

Keyes:
So what disturbs me is that an African American man who came from that legacy [of slavery] could do the same thing to dogs that white people did to us during the civil rights movement.

Dyson:
True. There's no question about that. But you know what? We're not dogs. We're not animals. We are African American human beings ... what he [Vick] did was reprehensible ... but to put dogs and animals parallel to black people is the extension of the legacy of slavery, not its contradiction.[5]

Keyes points out the ironic link between Vick's experience of racism and his brutality toward dogs, but Dyson brushes aside her comments and returns to race as the important issue at hand. Appearing on the *Late Show with David Letterman* in September 2008, comedian Chris Rock made the same move, saying that Michael Vick must be looking at pictures of Sarah Palin holding a bloody moose and thinking, "Why am I in jail?" Of course, Rock is right that there is something arbitrary about licensing hunting and criminalizing dogfighting. But Rock isn't animated by concern for animals here, he isn't exhorting us to take a more critical look at moose hunting. Instead he's foregrounding the issue of a racial double standard. In both of these examples, the focus on race subsumes, deflects, and ultimately denies the other set of moral claims being raised. Race, in other words, goes imperial in these instances, and the animal question is reduced to racism and silenced. Vick's defenders embrace human supremacy in the name of fighting white supremacy. When Kwame Abernathy called what was happening to Michael Vick an "electronic lynching," he of course echoed Clarence Thomas, who called the Senate Judiciary Committee hearings of 1991 a "high tech lynching," a rhetorical move which trumped the issue of whether Thomas had sexually harassed Anita Hill and secured him a seat on the US Supreme Court. In the Vick case, is the lynching metaphor being used once again to deflect our attention from another form of injustice—not male domination this time but human domination?

We have been at it for millennia, this project of animalizing animals or making them what we need them to be so that we can imagine ourselves the way we want to. According to Gary Steiner (2010), Plato and Aristotle's enshrinement of "reason" as the faculty marking superior beings was a critical development in this process, as was the Stoics' insistence that animals existed entirely for the sake of humans. The Bible (Old and New Testaments) indicated that humans were superior to other animals because they possessed a soul and that this justified their dominion (originally mild; after the fall, harsher) over the latter. Then there was René Descartes, writing in the seventeenth century. For Descartes, humans were defined by soul, mind, thought, and language: *cogito, ergo sum* (I think, therefore I am). Animals, on the other hand, were machines, pure matter, bodies unencumbered by souls or minds, and when they cried out upon being dissected while fully conscious, Descartes insisted that they were not feeling or expressing pain but making mechanical noises like a clock might make if certain mechanisms within it were triggered. By the time eighteenth-century utilitarian philosopher Jeremy Bentham pointed to animal sentience to argue for the moral considerability of animals—recall his famous quote, "The question is not, can they *reason*? Nor, can they *talk*? But, can they *suffer*?" (Bentham 1948)—he was swimming against the tide of Western theology and philosophy.

Despite the official line on animal subordination over the centuries, there were murmurings of subversion. For one thing, humans have always been good at maintaining contradictory ideas about animals and everything else (Ritvo 1998). Through the eighteenth and nineteenth centuries, Keith Thomas (1983) argues, various developments—including the articulation of less anthropocentric taxonomic systems such as the Linnaean one in *Systema Naturae* (1735–68); the emergence of urban middle-class pet-keeping; urbanization and industrialization and the distancing from rural life; and the expansion of fields such as astronomy, geology, botany, and zoology—all worked together to foreground human/animal continuities and generate new ways of relating to, thinking about, and feeling for animals. Animal advocacy emerged in Britain in the early 1800s in the person of William Wilberforce, notably a leading abolitionist *and* founder of the RSPCA, and soon after in the US, where abolitionism and animal advocacy were also closely tied. The publication of Darwin's *On the Origin of Species* in 1859, of course, gave a scientific boost to the ongoing reconsideration of the human/animal divide.

In the US, the last several decades have seen, simultaneously and contradictorily, an intensification in the commodification and instrumentalization of animals, driven by consumer demand and enabled by technological innovation, *and* a widening and deepening discussion over whether animals have the intrinsic right to be protected from exploitation. For animals, it is the worst of times and the best of times. "Animal capital," to use Nicole Shukin's (2009) term, is more salient than ever in the capitalist economy—from the unprecedented scale of factory farming to the manufacture of transgenic mice—but there is little doubt that our interest in the capacities and moral standing of animals is growing as well. With the intensification of mastery has come the intensification of doubt. If we have always been anxious anthropocentrists, in Erica Fudge's (2000) phrase, we have now become ambivalent ones as well—worrying not just about *how* to maintain domination over nonhuman animals, but also about whether we *should*. Genetic studies showing humans share 98.7 percent of their DNA with chimpanzees; ethological studies demonstrating the astonishing cognitive, emotional, and moral lives of animals; legal and philosophical arguments urging justice toward animals; investigative journalism exposing cruel farming and slaughtering practices; the emergence of the scholarly field of animal studies; the proliferation of animal law classes in leading US law schools; committed advocacy by animal welfare and rights groups; and yes, the extralegal practices of animal liberation groups—all of these developments together have initiated a cultural shift. The animal question has been raised.

Dogfighting is embedded in this contradictory picture. Public sentiment has turned decisively against it; what was once the sport of kings in England is now a felony in all 50 states. The Vick case furthered the criminalization of dogfighting, helping to make it a felony in states where it had been a misdemeanor and stiffening penalties. At the same time, according to law enforcement officials, there are indications that dogfighting in the US is actually growing not only in its traditional strongholds, white southern rural areas, but especially in urban black neighborhoods like Vick's (Burke n.d.; Mann 2007). Fighting dogs are showing up in Nike ads for basketball shoes and in rap—on the cover of DMX's *Grand Champ*, for example, and in music videos like Jay-Z's. There are more than 100 websites selling dogfighting equipment and dozens of journals chronicling dogfights (with the disclaimer that all accounts are fictional). An estimated 40,000 adults and 100,000 kids and teenagers in the US fight dogs (Peters 2008). With bets ranging from pocket change to tens of thousands of dollars, it is a half-billion dollar industry where gambling, illegal drugs, weapons, and animal cruelty explosively converge.

As Craig Forsyth and Rhonda Evans (1998) have shown, dogfighters deny charges of cruelty and insist that pitbulls choose to fight, that it's their nature, that they love it. They compare themselves to coaches and their dogs to prize-fighters. But what does it mean to say that dogs choose to fight? How do we assess free will and choice on the part of the dog in this context? Consider that in the world of dogfighting, dogs are genetically bred to alter normal canine behavior, which would cause the dog to growl in warning before attacking and to stop fighting when injured or when the other dog shows signs of submission. Consider that female dogs who resist breeding are placed in "rape racks" which immobilize them so that the male dog can penetrate them. Consider that from a tender age, dogs spend their entire lives outside, chained, just close enough to other dogs to keep them riled up but never close enough to have contact. Consider, too, that dogfighters continually taunt, starve, and drug dogs to heighten their aggression, that they use animals such as cats, small dogs, and rabbits (some of whom are stolen family pets) as "bait" to create an appetite for killing and for blood. Even after being bred and raised this way, some dogs still fail the viciousness test, showing little or no inclination to fight. These dogs are promptly killed.

Placed in a dirt pit ranging from 8 to 16 square feet and surrounded by a three-foot high fence, dogs are set upon each other for a fight to the death. Fights often last for hours and end with the death of one dog or when one dog cannot or will not continue. Dogs who do not die in the fight will often die hours or days later from broken bones, puncture wounds, blood loss, shock, dehydration, or infection. Because the activity is underground, dogfighters do not take their dogs to the vet, but either ignore their wounds or stitch them up themselves. Veteran criminal investigators describe themselves as shocked by dogfighting operations. They find pits full of blood, corpses of dogs, dogs with dozens of open wounds and half of their jaws missing, dogs with most of their bodies covered in scar tissue.

This account of a fight, taken from a dogfighting journal, gives you a sense of the violence involved:

> His face is a mass of deep cuts, as are his shoulders and neck. Both of his front legs have been broken, but Billy Bear isn't ready to quit. At the referee's signal, his master releases him and unable to support himself on his front legs, he slides on his chest across the blood and urine stained carpet, propelled by his good hind legs, toward to the opponent who rushes to meet him. Driven by instinct, intensive training and love for the owner who has brought him to this moment, Billy Bear drives himself painfully into the other dog's charge … less than 20 minutes later, rendered useless by the other dog, Billy Bear lies spent beside his master, his stomach constricted with pain. He turns his head back toward the ring, his eyes glazed searching for a last look at the other dog as [sic] receives a bullet in his brain. (Gibson 2005, 7–8 quoting C. M. Brown, "Pit," *Atlanta Magazine*, 1982, 66)

The most prized quality of a fighting dog is "gameness," or a willingness to fight to the death, because this is seen as reflecting the masculine strength of the owner. Alan Dundes's description of cockfighting as a "thinly disguised symbolic homoerotic masturbatory phallic duel, with the winner emasculating the loser through castration or feminization" (1994, 251) is a fine description of dogfighting, too. Rhonda Evans et al. write: "in the sport of dogfighting, the actual combatants serve as *symbols* of their respective owners, and therefore any character attributed to the dogs is also attributed to the men they represent" (1998, 832). The most despised dog is thus the so-called cur, the dog who turns away from the attacking opponent. Killing curs quickly and brutally helps to alleviate the owner's humiliation and restore his injured masculinity. The dog is put out of sight, but he or she was never visible to begin with. Human supremacy as expressed in dogfighting psychically and corporeally obliterates the dog.

Michael Vick saw his first dogfight at age eight but didn't get seriously involved until he was on the cusp of celebrity. The same month he was drafted by the Atlanta Falcons, he decided to start a dogfighting operation with friends and purchased the remote 15-acre property near Smithfield, Virginia for this purpose. His friends lived on the property and ran the operation and Vick visited every Tuesday (his day off) to oversee things. Vick and his friends used these methods to kill dogs whom they deemed insufficiently vicious:[6]

- Hanging by a nylon cord thrown over a 2 × 4 nailed between two trees
- Drowning by holding the dogs' heads submerged in a 5-gallon bucket of water
- Shooting in the head
- Electrocuting by attaching jumper cables to their ears and throwing them in the swimming pool
- Slamming repeatedly against the ground

The forensic report on the dogs whose bodies were found buried on Vick's property showed facial fractures, broken necks from hanging, broken legs and vertebrae, and severe bone bruising. Most had skull fractures, possibly from a hammer (Gorant 2010).

On ESPN's *First Take* in March 2013, commentator Stephen A. Smith was fired up that some of Vick's book tour appearances had been cancelled because of threats of violence. Vick was a "model citizen," "a quintessential role model," Smith remarked, as his colleagues called animal advocates "nuts" and "psycho" for their unwillingness to forgive. One quipped that Vick's book should have been called *Never Free* instead of *Finally Free*. Animal advocates are easy to ridicule. The social construction of them as morally out of joint, overzealous, and fanatical runs very deep in American culture. Seen through the lens of anthropocentrism, animal advocacy is by definition loony. Why don't these people pay attention to human issues? Homelessness? Child illiteracy? Juvenile diabetes? The presumption behind these questions is that humans come first, that they are more important than animals, yet the grounds for this argument are less firm than we might think.

The search for *the* single trait distinguishing all humans from all animals—what makes us uniquely human—has been as fruitless as it has been relentless over the past few millennia. Reason, language, self-consciousness, a sense of time and the future—all of these have been tried on and found lacking. Are humans more intelligent than other animals? Certainly, if we define intelligence as the particular types of cognitive skills which humans have and discount the intelligences of other species. But even here, as Peter Singer (2009) argues, the boundary fails once we grant that not every human has more of this kind of intelligence than every animal (a chimpanzee might do better than a mentally impaired human by some measures). When I first told my father, who is a first generation Korean American, that I was interested in animal issues, he asked why not work on the plight of North Korean refugees in China? For my father, the claim of blood, people, and nation takes priority over other claims. To me, his question was a reminder of the ultimate arbitrariness of choosing any cause to fight for. Who writes the rules governing love and commitment and where are they posted?

For some animal advocates, the Vick case was a chance to change America's mind about the pitbull. Pitbulls, in part because of their association with urban dogfighting, have gotten a bad rap. Indeed they have been racialized as urban/black/dangerous. As a result of this, pitbulls are dying. They are the dog of choice in US dogfights, and they are dying in dogfighting operations and in the fights themselves. They are dying because they are discarded in large numbers and end up in shelters where they are passed over for adoption and euthanized. Pitbulls now make up an estimated 30 to 60 percent of the dogs at shelters and are the most frequently euthanized type of dog in America (Muhammad 2012). Condemned as an entire breed, pitbulls are regulated and even banned by BDL, breed discriminatory legislation, in numerous localities across the nation. What dog rescuers want the public to know is that pitties are among the most loving family dogs, good with children, loyal, eager to please, and that while some are dangerous, through a combination of breeding and environment, most are not. The Vick case presented a rare chance for pitbulls to be seen as victims, to be seen as morally considerable and sympathetic and redeemable. Redemption was a central narrative in stories about the dogs taken from Vick's property (Gorant 2010). The Best Friends animal sanctuary in Kanab, Utah took in 22 of Vick's dogs—whom they have dubbed "the Vicktory dogs"—and invites the public to follow their progress as individuals struggling to overcome a difficult past. Some have gone to families with children and are thriving. Some will never leave the sanctuary because of their challenges.

The taxonomy of power which renders the animal object, slave, body, and commodity provides the ultimate justification for dogfighting, a practice in

2003. Greta Gaard's "Vegetarian Ecofeminism: A Review Essay" published in *Frontiers*.

which the dog's own needs, desires, interests are ignored. Of course, one can be anthropocentric and still oppose dogfighting. This is a position I suspect many people hold—dogs are only dogs but dogfighting is still wrong. But the supremacist, taxonomic thinking which holds that dogs are lesser beings is closely sutured with the thinking that holds that black people are lesser beings, that women are lesser beings, that gay people are lesser beings. Here I will go back and ask with Chris Rock—why is dogfighting wrong? What is the difference between dogfighting, on the one hand, and hunting, horseracing, Seaworld, the zoo, meat eating, and other institutionalized forms of animal usage? All of these involve commodifying other animals and reducing them to instruments of human ends, and all involve extreme measures of psychic and physical violence against them. This is what animal law professor Gary Francione (2009) meant when he intoned "We are all Michael Vick." Maybe seeing Vick's cruelty will open our eyes to our own cruelties. Maybe instead of making him into a monster, we will see ourselves in him or him in us. Maybe seeing the intimate entanglement of race and species, how race serves as a metric of animality, how the two taxonomies of power invigorate and reinforce one another, will push us to take another look at the daily practices which both systems of meanings authorize.

Animal advocates need to go beyond claims of colorblindness and racial innocence to grapple honestly with the persistence of race and the racial implications of their own labor. At the same time, they should not disavow animals who are being seriously harmed, whether it is racialized minorities doing the harm or whites. Michael Vick deserves a vigorous defense from racist practices but not a license to harm animals. Subordination in one sphere does not translate into moral immunity in another sphere. If it did, if we took this "get out of jail free" logic to its conclusion, then only white affluent straight men would be subject to critique and the rest of us would not.

It may be that forms of domination—white supremacy, heteropatriarchy, human supremacy, mastery over nature and more—are so intricately woven together, so dependent upon each other for sustenance, that they will stand or fall together. That as long as there are beasts, there will be Negro brutes. Can we imagine a world where white supremacy has been eradicated but human supremacy, heteropatriarchy, and the destruction of the planet motor on? Do we want to? Probably not, yet we remain, for the most part, in our separate silos, pursuing our separate struggles with hardly a sideways glance at each other. We embrace intersectionality as a theoretical insight, but do we accept what this might require of us politically? If Val Plumwood (1993, 2001) is right that it is the general posture of mastery, exploitation, and instrumentalization which has brought us to the brink of ecological destruction, how do we step back from that brink?

In closing, let me return to the American Dream. As a fable, the American Dream is stronger than ever, perhaps because under neoliberalism, the need for the fable grows as the real prospects for economic mobility shrink. But how is it that we have come to accept such a constricted and distorted definition of what it means to succeed as a human being? The American Dream is an individualist, materialist, consumerist, nationalist, exceptionalist fantasy with tremendous political potency. It seeks to create a citizenry that is so narrowly focused on personal enrichment that it doesn't see larger issues of economic polarization, poverty and hunger, war, the exploitation of animals, the despoliation of nature, global warming, the claims of other nations, the claims of other generations. Perhaps it is time to dream a new dream. To imagine the world we want to create and think about how to get there. To engage in a transformational politics where we rethink our identities, avoid that recursive loop of zero-sum competition, and focus our collective energies on challenging the architecture of supremacist thought and practice which imprisons all of us. To reimagine our relationships with each other, animals, and the earth outside of dominance. Justice in a multiracial, multispecies world. Perhaps it is time to dream a new dream.

Notes

1   This paper is adapted from a keynote address by the same name delivered on April 4, 2013 at the conference "Re-imagining the American Dream," sponsored by the Department of American Studies, University of Texas, Austin.

2   "Race Played Factor in Vick Coverage, Critics Say," Neal Conan's *Talk of the Nation,* National Public Radio, August 28, 2007. http://www.npr.org/templates/story/story.php?storyId=14000094. PETA Files. 2007. "Vick at the Office, Part 2." http://blog.peta.org/archives/2007/10/vick_at_the_off.php

3   "Richard Cebull, Montana Federal Judge, Admits Forwarding Racist Obama Email." http://www.huffingtonpost.com/2012/03/01/richard-cebull-judge-obama-racist-email_n_1312736.html. "Racist Obama Email: Marilyn Davenport Insists It Was Satire." http://www.huffingtonpost.com/2011/04/20/racist-obama-email-marilyn-davenport_n_851772.html

4   "The Great Yellow Hope." http://www.wbez.org/blog/alison-cuddy/2012-03-08/jeremy-lin-great-yellow-hope-97098

2005. Animal Abuse and Family Violence: Researching the Interrelationships of Abusive Power by Amy Fitzgerald published.

5   "'Supporting Our Own': Blacks Split on Michael Vick." National
    Public Radio, August 22, 2007.
6   "Bad Newz Kennels, Smithfield Virginia" Report of Investigation,
    Special Agent-in-Charge for Investigations Brian Haaser. August 28,
    2008. USDA Office of Inspector General-Investigations, Northeast
    Region, Beltsville, Maryland.

Ecofeminism and Veganism: Revisiting the Question of Universalism

Richard Twine

11

This article examines the complex question of universalism in the context of ecofeminist animal advocacy. This set of debates has broad importance for the "animal advocacy movement"[1] and is often exploited as a means by which to critique vegetarianism or veganism.[2] Given that ecofeminism already has a history of considering these debates, this article argues that intersectional ecofeminist thinking on this question has much of value to offer the broader animal advocacy movement which can add rigor to its liberatory roadmap.[3]

The subject of universalism and associated accusations of ethnocentrism could arise in a number of areas associated within either animal advocacy or ecofeminism, but it is the question of a universal vegetarianism or veganism[4] that has most often been the target of these discourses. The charge of universalism is sometimes assumed, by those making it, to instigate a wholesale denigration of vegan practice and to call into question the value of *anyone* being vegan. The charge could be seen partly as a defensive response to what is a threat to the very established social norm of animal consumption. However, at the same time, I do not wish to suggest that we can reduce all discourse critical

2005. After national attention beginning with a *Wall Street Journal* front page article, VINE becomes widely recognized for being the first to rehabilitate fighting roosters, an endeavor they explicitly characterize as an ecofeminist project.

of veganism to this—there *are* issues that need to be responded to. In the case of ecofeminism this has been part of the process of internal self-reflexive critique which accompanied the necessary, though drawn-out and sometimes paralyzing, debates on essentialism which were a feature of 1990s ecofeminism (Gaard 2011). Since ecofeminism has been such a threat to mainstream feminism (for example, contesting the feminist reliance upon humanist concepts of violence) it has—perhaps peculiarly and uniquely—developed in a highly self-reflexive way, constantly questioning itself in relation to critical themes of, especially, essentialism and universalism.

Dividing this article into four main sections, I begin by suggesting that, more than ever, the question of food practices and universalism is a pressing issue. I then review the various ways in which intersections between animals, nation, and racialization have occurred, before moving onto specific discussions by ecofeminists writing on the relationship between animal advocacy and universalism. I finish the article by noting differences between ecofeminists in their use of "contextual," different degrees of emphasis resulting in varying political visions. I suggest that intersectional approaches such as ecofeminism, which acknowledge the co-shaping of, for example, genders and species, and the dangerous tendency of analyses to reinscribe exclusions, have provided valuable tools and arguments for thinking through the complex terrain of food and universalism which is in stark contrast to the unfolding laissez-faire globalization of norms of high animal consumption.

Why universalism now?

This is, I suggest, a good time to revisit the question of veganism and universalism for several reasons. First, some popular proponents within the animal advocacy movement appear to assume that a universal veganism is a goal without first opening up that question to scrutiny. For example, Gary Francione's "The World is Vegan—if you want it" online campaign[5] is an attempt to inspire social change by reminding individuals that they have the choice to become vegan, which is presented as a practice of nonviolence. His website contains many translations of the campaign slogan so that activists from many countries can spread the word. While certainly laudable, Francione's campaign is open to criticism on several grounds. There is much diversity within veganism and it should not be assumed simply that it is always a choice for nonviolence. Human workers are also typically absent referents in the fetishization of food commodities and therefore, seen through an intersectional lens, vegan choices can certainly still be bound up in various forms of exploitation. In this way, politically, veganism is a beginning, not an end, because food and clothing choices are part of a broader economic system that exploits humans and animals alike. Choices are also socially constrained in all sorts of ways—be that through material means, access to knowledge, proficiency at skills, or adherence to social norms. Therefore, before we even begin to address the ethics of a vegan universalism, we are already faced with the need to be reflexive over what sort of change individual choices can bring and the recognition of sociological complexity in the realm of food practices.

Secondly, this is also a good time to revisit the universalism issue because, in the age of social media, campaigns against animal exploitation typically extend across national borders and so bring questions of culture and difference to the fore. There is now an unprecedented opportunity to know about, to opine, and to take action against practices taking place in areas of the world that are geographically distant. Significantly, contemporary communication technologies and new forms of mobility *also* further undermine the view of culture as a distinct bounded entity. Hegemonic *and* counter-hegemonic values are increasingly transcultural—for example, cultural forms, practices and identities inspire new socialities (such as the international animal advocacy or feminist movement) which transcend conventional state boundaries.

Thirdly, I argue, critical scrutiny toward a goal of universal veganism, though vitally important, takes on a tragicomic aura when one actually considers current trends around food practices and universalizing tendencies. It is thus important to remember (for vegans and those critiquing aspirations toward vegan universalism alike) to set debates around vegan universalism[6] within the larger context of the present-day universalization of Western food practices which include, of course, increasing global trajectories of meat and dairy consumption in, for example, Asia and Latin America. Although this does not detract from the importance of ecofeminist debates on contextual moral veganism and universalism, it is clear that such debates partly serve the purpose of teasing out an intersectional political vision rather than reflecting an impending empirical reality.

So specific discussions pertaining to a *theoretical* vegan universalism take place alongside the contemporary universalization of diets high in meat and dairy consumption, a trend also in need of urgent attention from an intersectional ecofeminist analysis. In terms of scale, contemporary accusations of food colonialism and imposition must then also be directed at the unsustainable Westernization of high rates of meat and dairy consumption in new parts of the world. However, it is important to add that not all facets of the universalism debate are theoretical. It is very much the case, as we shall see,

2006. The Food Empowerment Project founded, whose purpose is to create a more just and sustainable world by helping people recognize the power of their food choices.

that issues of racism and cultural difference can surface within veganism and animal advocacy and do reflect contemporary lived experience. These must continue to be addressed as part of the intersectionality vision of ecofeminism (see e.g. Gaard 2001; Harper 2010a; Kim 2007). Drawing on previous ecofeminist and intersectional scholarship that probes the universalism issue, including that of Claire Jean Kim, Greta Gaard and Marti Kheel, this article also intends to focus in on some of the key questions that are important to this vision.

An initial distinction can be made in the debates over ecofeminism and vegetarianism/veganism. The 1990s discourse that took place on listserves[7], books and journals (e.g. Adams 1990, 1993, 1994a, 1995b; Bailey 2007; Donovan 1995; Gaard and Gruen 1995; George 1994, 1995, 2000; Lucas 2005) often centered on either defining ecofeminism (Must ecofeminists be vegetarians?) or upon challenging feminism (Should feminists be vegetarians?). On this latter emphasis they were partly about the ecofeminist challenge to mainstream feminism over its human-centeredness or anthropocentrism. They *could* then be set in the context of previous critiques of (academic) feminism for centralizing gender and assuming a "woman" who was invariably white, middle-class, heterosexual, and omnivorous. Thus the ecofeminist challenge to feminism could itself be read as a challenge to the *universalism of feminism* (specifically a universal way of being human) and its failure to be intersectional *enough*, to name various relations (beyond gender), notably human/animal relations, as either exploitative, political, or worthy of attention.

These debates pertain to the way in which liberatory struggles often can end up reproducing new relations of power as they develop. It is notable then as the question broadens (from one of the vegetarianism/veganism of ecofeminists or feminists to that of arguing for vegetarianism/veganism across a society and across cultures) that the charge of universalism potentially amplifies. Could ecofeminists, or animal advocates generally, be seen as coming close to this in advancing a vegan universalism? As the example of Francione earlier illustrated, some animal advocates have argued for a universal cross-cultural veganism, though even here it is unlikely that vegan outreach in *every* geographical location on the planet is actually being pursued or suggested. Ecofeminists have been more explicitly cautious than the rhetoric of Francione's campaign, tending to introduce some sense of contextuality into their vegan ethics. Yet the implication has been from some (even in the culturally limited example of academic conference catering) that to advocate for vegan eating is ethnocentric, exclusionary, or even racist.

Intersections of animals, nation, and racialization

Before tackling that question, it is useful first to outline the various ways in which issues of animal exploitation or advocacy have been noted to intersect with questions of culture, nationalism, racism, and racialization.

In the case of national symbols and cuisines, animals are often drawn upon to construct a mythology of essential national character, perhaps emphasizing strength and masculinity, as in the case of the "British Bulldog" or the use of lions in British iconography. In the cultural political economy of property and exchange, animals as naturalized commodities become tied to nation or region, for example in the evocation of "*British* beef" for projects of economic growth. As Deckha points out, the deployment of meat for the performance of masculinities is inseparable from its intersection with nationalist and patriotic iconography (Deckha 2012, 139–40).

Ecofeminists (e.g. Adams 1994a; Twine 2001) have consistently argued that an important intersection between the exploitation of animals and racism is found in the way in which animals and notions of animality have been brought into the process of racialization, i.e. the marking out of "race." The animalization (to represent *as*, to compare *with*, other animals) of some people in the process of racialization and dehumanization is a longstanding trope that simultaneously constructed the racially unmarked category of "white" as "human" and "civilized." This has been drawn upon in various moments of colonialism, slavery, and genocide, with obvious examples including the enforced slavery of Africans, the British oppression of the Irish, and the Nazi construction of Jewish people. It remains part of the everyday repertoire of racist discourse today. As many have made clear, this use of human/animal dualism, as with its gendering, relies upon the anthropocentric assumption of animal inferiority, and so its comprehension is an important part of the wider intersectional understanding of power favored by ecofeminists.

I have already made reference to the ongoing universalization of a Western diet high in meat and dairy. The escalation of animal exploitation, the associated carbon footprint, locally imposed changes in land use, and the public health impact of this trend position it as a form of food colonialism, an unsustainable part of the global food system that is also wasteful of scarce resources such as water. It prolongs food insecurity and benefits Western transnational food corporations. There is no oversight to this change; it is typically naturalized, conceived as an inevitable part of "development" and the "free" market. Moreover it is even essentialized to the "human,"

2006. Maneesha Deckha's "The Salience of Species Difference for Feminist Theory" published in *Hastings Women's Law Journal*.

expressed in the view that (e.g. I have read this repeatedly in UN policy documents) when people have more disposable income it is assumed they want to eat meat—thus reinforcing a particular, historically white Western model of the "human." This underlines the way in which an interrogation of universalism is also closely tied to a politics around notions of human natures. For food corporations the relatively low(er) levels of animal consumption in many "developing" countries is understood as a growth and investment opportunity. These two examples of intersections are quite well known and clearly chime well with a politics of advocacy working against both animal exploitation *and* racism. They seem to unambiguously cast two social movements together as having shared aims and interests.

Of more concern to this article are those examples where animal advocacy might be seen to conflict with anti-racism. A particularly striking example of this is found in campaigning by far-right groups against halal and kosher slaughter. Better characterized as a disingenuous animal welfarism, groups such as the British National Party (BNP)[8] currently organize protests against ritual slaughter under the guise of concern for animals. In this way they attempt to tap into the discursive history of white civility constructed via a foil of racialized barbarity. Specifically for the BNP in this case, the premise is that Muslims are not "truly British." In the UK context animal advocacy groups have seen through this, arguing against *all* forms of slaughter. However the opportunity for an intersectional political alliance between pro-animal and anti-fascist groups is as yet unrealized.

This example is similar to that outlined by Claire Jean Kim in her case study of "immigrant" animal practices in California (Kim 2007). Instead of far-right groups exploiting animal welfare as a conduit for racialization, she outlines how the framings and targets of genuine animal advocates can become bound up in racist politics. Indeed, in a racist society, it is possible that pro-animal campaigns can garner broader support than they would otherwise when focused on the practices of minorities. Thus there could be cause for concern when advocates disproportionately target either immigrant practices or those in other countries. Kim however is particularly interested in how discourses of multiculturalism can narrow opportunities for critically opening up a discussion of animal ethics and I will return to her analysis later.

Potential intersections between animal advocacy and racism or colonialism also recall prior analyses of the Makah Whale Hunt (Gaard 2001; Hawkins 2001) as well as the recent analysis by Dinesh Wadiwel (2012) of the high-profile case of Australian opposition to the treatment of "Australian" live-exported cattle within Indonesian slaughterhouses. Wadiwel, rather than wanting to critique veganism *per se*, aims to make clearer the "racialized context within which veganism is practiced" (Wadiwel 2012, 1). Thus veganism and animal advocacy may be liable to become part of a process of constructing white civility when set within the context of a racialized geopolitics. A perverse proprietorial nationalism can be set in motion which intensifies an apparent concern for animals, here coded as *Australian* cattle under threat from racialized others. As Wadiwel argues, "the underlying message is that we kill our animals in a civilized way and *they* don't" (2012, 2), which is the similar rationale of the British National Party example discussed above, and arguably that of the welfarist and Western discourse of "humane" killing generally. In common with writers such as Harper (e.g. 2010a, 2010b) and Guthman (2008), Wadiwel wants alternative food discourse, animal advocacy, and vegan practice to be critically open to how it can act as a site for the reinscription of white privilege. Although veganism is certainly an attempt to reconceptualize the human away from habitual anthropocentrism, it is not itself immune from becoming entangled with exclusionary notions of the human, be they along lines of race or class. Such possibilities must then be part of the intersectional reflexivity that shapes analysis of the contexts of moral veganism.

Ecofeminists discussing animal advocacy and universalism

In the remainder of this article I explore in more depth these questions in reference to the key literature from those arguing from an ecofeminist or intersectional perspective. I already mentioned that ecofeminists have historically been cautious toward universalism. This prudence was shaped by the ecofeminist critique of both utilitarian and rights-based theories of animal ethics (e.g. see Donovan and Adams 1996). That traditional ethical theory had a tendency to be pronounced universalistically has been part of a broader critique that was further suspicious of ethics divorced from everyday life *and* of the disavowal of emotionality. For example, Marti Kheel specifically associated universalism with her understanding of masculinity (Kheel 2008, 3). Thus universalism became perceived as something to avoid in favor of sensitivity to the lived context of differently positioned people.

An early contribution to the ecofeminist literature on ethics of care and contextual moral vegetarianism[9] by Deane Curtin argued against an absolute moral rule prohibiting meat eating under all circumstances (Curtin 1991, 69). For Curtin, personal contextual relations (essentially unlikely emergency situations) and geographical contexts may also be relevant where a choice to

2006. Film-maker Tami Wilson's *Flesh* premieres, examining the connections between women and meat in a society obsessed with flesh.

subsist without killing animals could be seen as constrained by ecology and climate. In spite of these contexts Curtin's vegetarianism is, I would argue, not far from universalistic—in other words, under most circumstances his ecofeminist ethic of care argues against the consumption of animals. During the 1990s and early twenty-first century the debate over contextual moral vegetarianism would be taken up by people such as Karen Warren, Val Plumwood, Greta Gaard, and Carol Adams.

Before noting some of these arguments I wish to first focus on the more recent work of Claire Jean Kim, not an ecofeminist as such but someone who does critical work around the intersection of race and species. In her case study of some "immigrant"[10] animal practices in California she focuses on controversies involving the practices of both horse tripping in Mexican *charreadas* (rodeos) and the practices of live animal market workers in San Francisco's Chinatown. While very aware that such practices can be exploited to enhance the representation of the otherness of immigrant communities, Kim is nevertheless critical of what she sees as a simplistic condemnation of animal advocacy opposition to them as racist. She is also skeptical of defenses of these practices in terms of them being traditions that are an integral part of communities. More specifically her critique is centered on the way in which multiculturalism is an anthropocentric frame and, for her, is deployed in these debates in an imperialistic manner. Through a close reading of the legal challenges to these practices she argues that:

> Defenders of immigrant animal practices have reacted by evoking a multiculturalist interpretive framework that situates these interventions about animals in a long history of White aggression against powerless immigrants of color. In this view, the majority is guilty of judging immigrant minority cultures to be deficient and wrong (ethnocentrism), seeking to impose dominant cultural values on marginalized cultures (cultural imperialism), and enunciating an impassable racial difference between the "self" and the "other" in order to bolster ongoing efforts to exclude the latter from meaningful membership in this nation (racism and nativism). Animal advocates, it is said, operate with a "double standard," targeting foreign-born minorities while ignoring the majority's own cruel animal practices. (Kim 2007, 2)

Whilst some advocates may indeed operate such a double standard, Kim accuses the defensive position as working to close off critical ethical questions around the treatment of other animals. In one respect her critique goes to the heart of both the anthropocentrism of multiculturalism and a problem with the argument of "cultural difference" as a defense of animal exploitation. Those arguing from an intersectional or ecofeminist perspective cannot simply take the concept of "culture" as read. The ethical tradition of ecofeminism has been to question the production of a discourse of culture as exclusively human. Thus we must note that the "multi" of multiculturalism is insufficiently multiple and the terrain of "cultural difference" must be broader than the human. While "culture" will continue to be "too much" of a term for many to apply to other animals,[11] even without this designation what we share with other animals both socially and corporeally ought to be enough to transgress the human/animal dualism of moral considerability.

Two further points by Kim (2007, 7) are relevant to enriching the debate on universalism and cultural difference. The deployment of a multiculturalist framework to defend exploitative practices against other animals not only essentializes cultures as purely human but both majority and minority cultures are represented as coherent unitary wholes (a similar point is made by Gaard 2001). Appeals to respect other cultures risk conflating a given practice with the entirety of a culture and of reifying that culture as static. Secondly, it "elides ideological, rhetorical and strategic differences among animal advocates, the mainstream media, politicians and others, by lumping these distinct entities together as 'the dominant groups'" (Kim 2007, 7). This is an important point since animal advocacy cannot be said to represent a dominant part of any Western culture. Even in its non-vegan welfarist form it is not so. Moreover vegetarianism is clearly in significant respects also a non-Western practice (Kheel 2004, 335; Spencer 2000) even though we must acknowledge the cultural diversity of reasons for vegetarianism or veganism. Animal advocacy as a social movement constitutes an oppositional politics, arguing for a radical revision of dominant Western cultural practices and institutions. While this does not preclude the possibility that an alternative and minority (pro-animal) politics could act oppressively toward, for example, immigrant groups or other cultures, it renders problematic attempts to situate them within a history of dominant Western cultural imperialism. Recognizing this arguably aids the possibility of intersectional coalition, since both animal advocates and discriminated communities have a shared interest in opposing a capitalism that instrumentalizes according to constructions of race and species via intertwined legacies of animalization and racialization.

Kim also gives a clear and specific stance on the question of universalism. She reminds us that "claims about the culture-boundedness of all assertions are nonsensical in that they themselves assume universal form" (2007, 12). Marti Kheel makes a similar point, stating that "the view that all advocacy of vegetarianism as an ideal is inherently imperialist is an unjustified universal claim of its own" (Kheel 2008, 267n. 159), even if this could also be said to contradict her conflation of universalism with masculinity. Perhaps then we need to be more nuanced about assertions that all instances of universalism are, for example, ethnocentric, imperialist, or masculinist. Kim takes the view that the recognition of a small number of universal values is morally necessary[12] (Kim 2010, 59), proposing that the prohibition of cruelty (against both humans and animals) ought to be one such universal value (Kim 2007, 12). Again human/animal dualism shapes moral considerability when using a discourse of "cruelty" or "violence" which entails that, in some sense, the exploitation of other animals is typically taken less seriously. This is not to say that human/animal dualism exists in exactly the same form in all human cultures. However anthropocentrism (like patriarchy) does seem to be significantly cross-cultural,[13] even if its history, trajectory, and experience exhibit cultural specificity.

Kim's thoughts resonate with the earlier important work of Gaard. Both advocate for forms of dialogue as inescapable strategies for addressing such cross-cultural controversies—a point I return to later. Gaard's (2001) focus was on the Makah Whale Hunt, a controversy that came to prominence in the late twentieth century over the right or not of the Makah tribe in North West Washington State (USA) to resume whale hunting. In common with Kim's case studies, the situation of the Makah also reflected a marginalized group long subject to racism. However a relevant difference here in the practice was that the Makah had not hunted whales since 1913 and therefore were seeking to *resume* a practice. Furthermore, that the resumption was not predicated on reasons of subsistence meant that it did not satisfy the International Whaling Commission's (IWC) criteria for "aboriginal whaling" (Gaard 2001, 5–6). Nevertheless the US government managed to gain permission from the IWC for the Makah to kill whales between 1998 and 2002, with the first animal killed in May 1999. The pro-hunting claims of (in fact, a segment of) the Makah also came up against the problem that broader cultural sensitivity, the regulatory environment, and scientific knowledge of animals such as whales have shifted considerably during the past 100 years.

Both Kim and Gaard are sensitive to the specific historical context of communities attempting to survive in the face of racism and economic marginalization. At the same time, both want to open an ethical space for the considerability of other animals in the face of the simplistic denouncement of animal advocacy as neo-colonialist. Moreover both analyses are critical of tendencies to homogenize such communities. As Gaard points out, in the case of the Makah there was significant opposition to the whaling *amongst* the community, with some evidence of a traditional gendering of opinion vis-à-vis the advocacy of whaling as a method for Makah cultural renewal. She argues that "In the case of the Makah, the whale hunting practices of a certain elite group of men have been conflated with the practices, and substituted for the identity, of an entire culture" (Gaard 2001, 17). There is certainly a responsibility for ecofeminists not to discount an analysis of gender or other divisions within a culture or community in order to better grasp the context of why certain practices are pursued. Approaching communities as undifferentiated, as homogeneous, only reinforces their objectification (see Plumwood 1993). Moreover this also constitutes an impoverished sociology and emphasizes the need for intersectional approaches such as ecofeminism to conduct grounded empirical research in order to properly grasp the complexity of a controversy. Although ecofeminist ethics began by distancing themselves from abstraction, analyses such as Gaard's and Kim's suggest that ecofeminism is enriched by sociological and historical analysis. Intersectionality is not something one can hope to understand only as a disengaged philosopher. Similar arguments have been made within feminist bioethics where many have also critiqued philosophical over-abstraction in ethical analysis (see Twine 2010a).

In spite of the similarities of approach between Gaard and Kim, Gaard is more reticent. For example, she does not explicitly advocate for a limited sense of universalism and states in relation to the Makah controversy that "In this specific intersection of historical, political, and cultural contexts, it is not the place of non-native feminists and ecofeminists to challenge even what we perceive to be oppressive features of marginalized cultures; rather only members of a specific culture are positioned to lead an inquiry into tradi-tional cultural practices" (Gaard 2001, 18). Such a principle raises difficult questions since it may be the case that there are *no* members of a community that want to lead an inquiry, especially if it could be read as a form of betrayal. Here we can read "lead" literally and assume that Gaard does not mean that one has to be a member of a specific culture in order to voice criticism. She advocates for a postcolonial ecofeminist perspective where the responsibility is on ecofeminists to learn about the particular culture and context related to the practices in question. This further highlights the need for embedded and proximal methodologies; Gaard argues that people who can act as "border crossers" between different perspectives and spaces are vital for setting up mutually respective dialogue (Gaard 2001, 19–22).

Perhaps then this is not so far from Kim's take on such conflicts. Although Gaard does not argue for a limited universalism she does state, for example, that ecofeminism pursues "certain minimum conditions for ethical behaviour" and wants to support certain "basic principles" (Gaard 2001, 3; see also Kao 2010, 626). Kim also supports a similar mutually sensitive dialogue in which both sides open themselves to scrutiny. In the cases noted in this article such as in California, Washington State or Australia, and elsewhere, majority animal practices also have to be open to critique. Their scale and

2006. Maine becomes the first state to pass a law that allows companion animals to be included in orders of protection.

effects are, after all, typically more systemic. Kim concludes by saying that "Immigrants need protection from cultural imperialism and nativism, but receiving and giving moral criticism and engaging others on issues of moral concern is an important part of membership in a moral community. The risks of being seen as outside of this community may well be higher than the risks of being included" (Kim 2007, 14). Both Gaard and Kim wish to construct a way forward that is accountable to the intersections of different relations of power, but also push intersectionality as a concept beyond its traditional anthropocentric usage (see Twine 2010b). Ecofeminism, as I have argued, can realize such analyses as long as it strives for methodologies and politics that account for the complexity of social, cultural, and historical context, so that we might better understand how certain practices begin to take hold.

Contesting "contextual"

To end this article I want to examine the uses of "contextual" by ecofeminists a little more closely. "Contextual" has been used in two main ways, first to stress the importance of understanding the contexts of practices and secondly to actually highlight specific contexts within which animal use could be deemed unavoidable. In particular we can note this in differences of perspective over the second sense between Val Plumwood and her critique of some ecofeminists, notably Carol J. Adams and Marti Kheel. I focus mostly on Kheel given her specific position on the question of universalism. Although a committed vegan, Kheel took care to avoid advocating for vegan universalism. As previously noted, in tension with Kim, Kheel viewed universalism as masculinist. Shaped by this and her grounding in ecofeminism, she was wary of abstract norms and universal rules. Instead Kheel argued for what she referred to as an "invitational approach" (2004, 328) to vegetarianism[14]. Kheel wanted ecofeminism to focus on deconstructing normalized practices of animal consumption, to place the onus of justification there, in order to "clear a space in which to plant the seeds that invite the vegetarian ideal" (Kheel 2004, 329). Yet it is important to grasp the complexity of Kheel's position. It is inaccurate to portray her simply as a contextual ecofeminist countering universalism. In fact she was clearly critical of some contextual ecofeminists (Kheel 2004, 334).

As we saw earlier, Curtin's approach was to account for contexts of emergency and geography as potential examples of unavoidable animal use. However, others such as Val Plumwood argued for a broader understanding of "contextual" and espoused an anti-vegan stance (see Plumwood 2000, 2003,

2004). Plumwood's position instead was "semi-vegetarian" and was against factory farming rather than killing animals per se (Plumwood 2004, 53). She correctly wanted the "human" to be resituated in ecological terms but incorrectly framed veganism as precluding that possibility (see Plumwood 2003, 2). Just because vegans do not eat animal products does not mean that everything else they do eat does not somehow embed the (vegan) human in the rest of nature. Nor do vegans, as she claims, insist that neither humans nor animals should ever be conceived as edible (Plumwood 2003, 2). A radical reorganization of human death and burial would be one way, for example, to promote the ecological benefits of human edibility. Human predation itself (preying upon, or being live prey) is hardly a prerequisite for ecological flourishing.

In a straw man critique of the work of writers such as Carol Adams and Marti Kheel, Plumwood portrayed vegan ecofeminists as being universalist and of being anti-predation generally. Although Plumwood made useful points over choice—Western urban vegans should not assume that their choice of foods are reflected in other parts of the world (Plumwood 2004, 306)—her latter work unhelpfully and unnecessarily re-emphasized a divide between environmentalists and animal advocates. While vegans such as Kheel and Adams were keen to underline that predation had been over-emphasized and used to naturalize meat consumption in humans, Plumwood conflated this with a view that denied predation generally in other animals. Furthermore there was a paucity of evidence to suggest the sort of accusation of universalism that Plumwood was making had validity.[15]

Kheel deflected such charges by distinguishing her invitational approach from one that would seek to *impose* practices on other cultures. Here veganism is an ideal presented as a positive mode of living more caring and empathic lives (Kheel 2008, 246) that can be offered to people *alongside* a critical incitement: to reflect upon the arbitrariness and damage done by norms and expectations of animal consumption. In arguing against some versions of contextual vegetarianism, she argued:

> Typically, the contextual approach focuses on the importance of understanding and respecting meat eating within the overall context of particular cultures, without examining the sub-cultural contexts that exist within the larger culture. While it is important to try to understand, and where appropriate, respect the practices of other cultures, this should not preclude a deeper analysis of the cultural associations that may underlie those practices, and in particular the cultural associations between masculine self-identity and meat eating. (Kheel 2004, 335)

This concurs with the view of both Gaard and Kim, that it is a poor analysis which fails to try and understand social differentiation within a particular culture or community. In her book *Nature Ethics* Kheel proposed that advocating for veganism is fully consistent with an ecofeminist contextual approach (Kheel 2008, 235). She explicitly did not construct a rational argument for veganism, arguing that it is also the case that people do not typically evoke such arguments for why they do not kill other *humans* (Kheel 2008, 235). This highlighted her interest in social norms. There is little public discourse on rationally justifying not killing other humans because it is a taken-for-granted social norm, a keystone of the social order. Animal consumption and anthropocentrism operate similarly; that humans eat and kill other animals is as normal as the understanding that we do *not* kill other humans. Acutely aware that veganism is anything but a social norm, Kheel purposively avoids dictating veganism as a universal injunction because she wants her *primary* focus to be on radically troubling the norm of animal consumption which operates, for Kheel, to guarantee society's access to animals' bodies (Kheel 2008, 236). Although some may read this move as a retreat from the "dangerous territory" of universalism, Kheel's tactic here is consistent, I suggest, with what I earlier referred to as the tragicomic character of the debate around vegan universalism.[16] Social norms are not static, but there is no impending probability of vegan universalism. Since ethnocentric patterns and trends of high animal consumption associated with specific developments within twentieth-century Western culture *are* currently universalizing to other parts of the world, it does seem fair that it is *these* practices, and those interests that support them, that must be foremostly called to account. Furthermore any realistic examination of food practices being culturally imposed must engage with the mass prescription of animal consumption as a norm in the majority of human cultures *and* attend to the historical, social, and economic reasons for that. On the one hand, high rates of meat/dairy consumption are an increasingly globalized practice, and yet specific historical configurations of organizing this practice (see Shove and Pantzar, 2005, 57–8) are important for considering both its longevity as well as prospects for alternative practices to take hold.

In this chapter I have set the question of (ecofeminist) vegan universalism in a broader context.[17] The emergence of ecofeminist vegan ethics can be situated within the wider relationship between ecofeminism and other types of animal ethics and its challenge to the anthropocentrism of mainstream feminist thought. Any attempt to advocate for large-scale changes in eating practices cannot subsist alone upon ethics and must acquaint itself with the sociological, historical, and cultural dimensions of eating and human/animal relations.

While the shaping of ecofeminist thought has attuned itself to thinking about tensions with universalism, debates on *vegan* universalism are significantly overshadowed (though not trivialized) by the economic and cultural globalization of high rates of animal consumption. What if more critical discourse was directed at this significant cross-cultural change as being enmeshed within ethnocentrism and colonialism?

To be clear, I have not used this point to diminish the importance of also directing attention at the ways in which pro-animal politics can become bound up in new or old exclusions, new constructions of whiteness, and how they can become enrolled by racists partaking in "opportunistic animal advocacy." This is the task of posthumanist forms of intersectionality to be reflexive to precisely such developments. Although beyond the scope of this article, I have suggested that there is a need to more carefully consider research methodologies that can appropriately be positioned to understand the various contexts that shape the emergence and durability of human/animal relations—in effect, a more coherent approach of how to practice intersectionality.

Broadly four positions emerge from the ecofeminist or intersectional literature on the specific question of animal advocacy and universalism. Val Plumwood has argued for a semi-vegetarianism which is more an opposition to factory farming rather than to meat consumption per se. I argued that her critique of veganism fell short in several ways and that her perception of an uncritical universalism in writers such as Carol Adams and Marti Kheel was misplaced. Deane Curtin, and arguably most ecofeminists, have favored a form of contextual moral veganism which I have characterized as a "near universalism"; it certainly implies that in most contexts humans can and should be vegan. I read Claire Jean Kim's position as similar but with a limited model of universal values based especially around opposition to cruelty. Marti Kheel's somewhat different approach was to construct an "invitational approach" to veganism and to focus on the compulsory nature of norms of animal consumption as universals in need to critique. Such positions will continue to offer the wider animal advocacy movement important tools for a liberatory roadmap. I have also suggested that ecofeminists have an interest in not accepting simple defenses of particular human/animal relations based on (human) cultural differences precisely because they have had a stake in not assuming an anthropocentric and atemporal conception of "culture." To make the argument from "culture" for a practice such as bullfighting is to use a speciesist term and frame in order to perpetuate a speciesist relationship. The political crux of ecofeminism and kindred accounts of intersectionality is to not only create cultures in which other animals matter, but to move "culture," precisely, away from norms of animal exploitation.

Notes

1   I do not wish to present this as either homogeneous or tied to any particular geography.

2   I do not use the term "animal rights movement" since I prefer to stick to a theoretically correct description. Given the diversity of political and ethical thought, one does not, strictly speaking, have to commit to a rights view to advocate for other animals. From an ecofeminist perspective it is also worth stating that there can be no such thing as an isolated "animal advocacy movement" since the "question of the animal" intersects with intrahuman relations of power. Consequently veganism can only ever be one, albeit important, part of an integrated ecofeminist political practice.

3   I see this chapter as related to my recent paper on domestication— see Twine, R. (2013) "Is Biotechnology Deconstructing Animal Domestication? Movements toward 'Liberation'," *Configurations* vol. 21, no. 2, 135–58. Their relatedness consists in an attempt to think critically about animal liberation, and to probe what the "liberation" of other animals can actually mean. The *Configurations* paper specifically considers what liberation can mean in the context of domestication. Together with the issue of universalism discussed in this chapter I see these as some of the most challenging questions for the rigor of the animal advocacy movement.

4   During the 1990s there were intensive debates within ecofeminism on the question of whether eco/feminists should be vegetarians, as well as around the notion of contextual vegetarianism as mentioned in the Groundwork in this volume. Notably the term "vegetarianism" was mostly used instead of "veganism." I would contend three reasons for this. First, I expect some North American writers used vegetarianism but meant veganism. Secondly, since the 1990s I would argue that vegetarianism has lost a lot of credibility as a consistent ethical position within the animal advocacy movement but at *that* time it was still deemed credible. Thirdly, and relatedly, during the first decade of the twenty-first century, notably in Western countries, there has been an ethical shift toward, and cultural normalization of, veganism as the preferred and more consistent practice of animal advocates. So much so, that an ecofeminist arguing today for ovo-lacto vegetarianism would suffer from a credibility problem.

5   See http://www.abolitionistapproach.com

6   Be they critiques of veganism that assume that a universalism is being proffered, or the important ecofeminist debates on universalism.

7   I certainly remember participating in a heated debate in the mid-1990s on the "ecofem" listserv based on the University of Colorado server on the subject of whether ecofeminists should be vegetarian. It now seems quaintly naïve in retrospect (see note 4).

8   See http://www.bnp.org.uk/news/regional/halal-protest-sunderland-25th-aug-2012

9   See note 4.

10 This is not intended as a racist conflation of the term "immigrant" with anyone not classified as "white."

11 The complexity of this question is beyond the scope of this article. Needless to say that many arguments have been made for animal culture, and that human and animal cultures are hardly distinct from each other, which calls into question the assumption of culture as purely human.

12 Kim's thinking on universalism is influenced by the theorist of multi-culturalism Bhiku Parekh.

13 I accept that the term cross-cultural could be deemed problematic for the way in which it could be read as assuming that a singular culture is a clearly demarcated entity.

14 Kheel is an example of a writer who used the term "vegetarian" to in fact mean "vegan." See her explanation (Kheel 2004, 338n. 1).

15 For a well-argued response to Val Plumwood's critique of Adams and Kheel, see Eaton 2002.

16 Kheel is very clear on her direct thoughts on universalism so I do not think she can be accused of being evasive. Her stance is not so different to that of Deane Curtin. For example, she stated that "There is nothing inherently oppressive in encouraging vegetarianism or veganism as ideals, while recognizing that there may be environmental and climatic factors that make them difficult in some cultures. Advocating ideals is not the same as seeking to impose one's beliefs on other people and other cultures" (Kheel 2008, 236).

17 My focus has largely been on animals exploited for food. For a broader philosophical discussion of animal ethics and context, see Palmer 2010.

2006. Anna Lappé and Bryant Terry publish *Grub: Ideas for an Urban Organic Kitchen.*

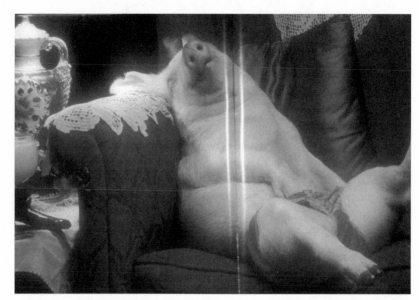

Figure 12.1 "Ursula Hamdress" from *Playboar*. This image appeared in *The Beast: The Magazine that Bites Back*, 10 (Summer 1981), 18–19. It was photographed by animal advocate Jim Mason at the Iowa State Fair where it appeared as a "pinup." (More recent issues of *Playboar* have renamed "Ursula" "Taffy Lovely").

Why a Pig?

A Reclining Nude Reveals the Intersections of Race, Sex, Slavery, and Species

Carol J. Adams

12

Several recent books of critical theory have included inves-tigations of the traditional Western depiction of women's beauty or women's sexual availability especially as it has been captured in the "reclining nude" pose. Yet, as critical theory traces these depictions forward into the late twentieth century, it has failed to recognize specifically how this tradition has leapt the human body. In this essay, I return to an image that has been vexing me for 30 years, and illuminate it as an example of the reclining nude as it was imposed on an other-than-human body. I argue that cultural theory must include consideration about species hierarchies and attitudes when examining racial and sexual representations. Otherwise it is impoverished. Attitudes about sex and race that continue to be imposed on an other-than-human body are permitted to be retrograde and oppressive, escaping the kind of scrutiny that would be brought to bear if the representation were imposed on a human body. I seek to change this.[1] In specific, I am concerned about the image of "Ursula Hamdress" (see figure 12:1). I first encountered this image in the

2007. Wake Up Weekend begun in Grand Rapids, Michigan—an annual grass-roots celebration of animal-friendly advocacy, art, food, education, music, philosophy, and religion that seeks to solicit the many but often hidden ways in which the exploitation of animals sustains other entrenched forms of oppression that we all stand against: global hunger, diseases of affluence, the exploitation of women, racial injustice, and environmental degradation.

early 1980s. Jim Mason had been in the Midwest and decided to go to the March 1981 National Pork Producers' conference in Kansas City, Missouri. He wandered through the trade show, flush with 480 exhibitors and 15,000 visitors. The implements of contemporary hog raising are there: farrowing stalls (which imprison a sow who has just given birth), cages, pens, slatted floors, feeders, etc. "Here Hess & Clark, Pfizer, Dow Chemical, Elanco, American Cyanamid, etc. are exhibiting the myriad varieties of antiobiotics, disinfectants, growth promotants, pre-mixes and other factory drugs and supplies." Jim, who by that point had many visits to factory farms under his belt, notices the absence of dust, manure smell, dank, acrid air, and "no shrieks of crowded pigs."

He heard a buzz and saw lots of people gathered around something; they were all staring at and talking about the picture of *Ursula Hamdress*: "a photograph of a pig in panties sprawled in a chair. I overhear one of the men explain how a veterinarian sedated the pig so that she would hold still for the picture" (68).

"Ursula" was named for actress Ursula Andress, a "sex symbol" (in the terminology of the 1960s) as a result of her role in an early James Bond film. In 1965, she posed for *Playboy*. "Ursula Hamdress" with her painted trotter nails and red Victoria's Secret-like panties, was posed as a centerfold for the magazine *Playboar—the Pig Farmer's Playboy*.

In the early 1980s, I recognized that two "genealogies" had combined in Ursula Andress—the pornographic and the domesticated farm animal. In terms of the pornographic, "Ursula Hamdress" was posed as though she were the centerfold for a pornographic magazine. The accoutrements in the photo-graph were staged in such a way as to evoke a nineteenth-century brothel, but the being was distinctly different than the kind of being usually found in that setting or as a centerfold: a pig.

The other genealogy—that of the lives of farmed animals, from which "Ursula" had been elevated into a human-inspired environment—had its own setting and accoutrements. Since the publication of the 1980 book Jim Mason's and Peter Singer's *Animal Factories*, with its photographs and text, an animal activist who encountered the photograph of "Ursula Hamdress" would recognize something distinctly different about her, too: she showed none of the signs of having lived the kind of life the majority of her sister sows endured. She was unblemished. No other sow had chewed on her, forced into a cannibalism through the stress of crowded conditions.

By the time the 1990s rolled around, Ursula Andress's iconic sex symbol status had faded. The "joke"—Ursula Andress in *Playboy* becomes "Ursula Hamdress" in *Playboar*—had lost its referent for most consumers. However the visual referent remained. *Playboar* circumvented this dated association by making just one change: they changed the name of the pig. "Ursula Hamdress" became "Taffy Lovely." The sole updating they did for this magazine of barnyard/school/fraternity humor—the humor derived from dominance—was to change the name of the sedated pig.

Otherwise, that issue of the magazine has stayed exactly the same. All of the other visual and verbal jokes and double entendres were left untouched and it seamlessly moved into the twenty-first century with the iconography of the twentieth. To the editors of *Playboar,* there was a sense of an unchanging set of consumers from the 1960s to the twenty-first century, and they weren't just pig farmers.

The genealogy of the reclining nude

Another genealogy exists into which "Ursula Hamdress" fits—a genealogy identified recently by several important cultural critics, but a genealogy that in their discussion does not include "Ursula." I will argue that it should.

The three works I will consider are David Harvey's *The Conditions of Postmodernity* (1997) which describes the movement from modernity with its emphasis on rights and a teleology of progress to postmodernism with its flowering of fluidity, multiplicity, plurality; Michael Harris's *Colored Pictures: Race and Visual Representation* (2003) which examines black artists' response to racist imagery; and Nell Painter, who makes the invisible and the universal perspective associated with whiteness in Western culture visible and specific by providing *A History of White People* (2010). Each author devotes visual space to the evolution of the pose called "the reclining nude." Scholars concerned with race-making, representation, and the transition from modernity to postmodernity all gravitate to images that are the precursors of and motivators for the posing of a (possibly dead) pig. Placing this (possibly dead) pig within the larger cultural tradition they analyze is important.

David Harvey's *The Conditions of Postmodernity (1997)*

In Chapter 3 of *The Condition of Postmodernity*, David Harvey takes a big risk. He chooses five images that feature naked women without alluding to that specific fact. He begins by introducing a chart by Hassan that identifies the "Schematic differences between modernism and postmodernism" (43). It

2009. Bryant Terry's *Vegan Soul Kitchen: Fresh, Healthy, and Creative African-American Cuisine* published. He summarizes his approach to the issues of health, food, and farming as "Start with the visceral, move to the intellectual, and end with the political."

is as though postmodern binaries do and don't exist; binaries are a remnant of the modern project; but these schematic differences can't be freighted with meanings of essentialism or universalism. Here is a sample of the binaries he and Hassan identify:

Modernism	*postmodernism*
Form	antiform (disjunctive, open)
Design	chance
Hierarchy	anarchy
Distance	participation
Centering	dispersal
Semantics	rhetoric
Signified	signifier
Metaphysics	irony

Race, sex, and species are not "schematic differences" for Hassan. After 12 pages of discussion of these binaries among other things—12 pages in which gender is neither theorized nor examined—one turns the page and finds David Salle's *Tight as Houses* (1980). Salle's is the only image that is allowed to take up the space of an entire page. It is difficult to "read" in its detail—there is a sketch (written upon the negative of a photograph?) that is imposed over a photograph of a woman's well-rounded naked body. It seems to illustrate Harvey's statement that "the deconstructionist impulse is to look inside one text for another, dissolve one text into another, or build one text into another." In Harvey's words, "the collision and superimposition of different ontological worlds is a major characteristic of postmodern art" (50). The image is the site of collisions, of building a text upon another text. That this text is a woman's body is both obvious and untheorized.

Four pages later, naked women begin to appear more frequently. Titian's *Venus d'Urbino* greets us with her unabashed stare and her hand coyly covering her pubic area. This *Venus*, "one of the first reclining nudes in Western art" and originally called by its owner *"la donna nuda*—the naked woman" (Harris, 128), has generated many successors, including, as Harvey notes (twice!), Manet's *Olympia*. Turning the page, there she is, assured, brash even, her eyes meeting ours, her hand laying on top of her right leg and thus, also, obscuring her pubic area.

Across from *Olympia,* Rauschenberg's *Persimmon* stares at her viewers through a mirror. Harvey remarks that "Rauschenberg's pioneering postmodernist work *Persimmon* (1964), collages many themes including direct reproduction of Rubens's *Venus at her toilet.*" Juxtaposed with *Olympia, Persimmon* appears to rework Manet and Titian as well—what faces out now faces in, but the mirror restores the gaze. According to Robert Hughes's *The Shock of the New*, Rauschenberg "was fond of embedding an ironic lechery in his images" (335).

In *his* choice of images, Harvey appears to have had some fun evoking an ironic lechery. At the end of the chapter, a naked young woman stares at us from an advertisement for Citizen Watches. She recalls *Persimmon,* with her backside facing frontward in the advertisement, naked except for a watch. As I noted in my use of images in *The Pornography of Meat*, watches are a primary vehicle for inscribing dominant attitudes (57). In postmodern times, advertisement strategies influence art and art influences advertisements—superimposing images and creating referentiality across their once discrete fields.

Harvey's decision to use paintings that represented women's bodies to illustrate the evolution of art into postmodernity seemed to backfire on him. After the first edition of his book was published, Harvey's genealogy of images quickly came under critique. Tellingly, feminists found both acts of commission and omission.

One of the commentators was D. Massey, who wrote in "Flexible Sexism" a stunning critique of the implicit sexism in his analysis that ignores any insights from feminist theory and the explicit sexism in the illustrations in Chapter 3. She writes:

> His commentaries ponder the superimposition of ontologically different worlds, or the difference between Manet and Rauschenberg, but they are oblivious to what is being represented, how it is being represented and from whose point of view, and the political effects of such representations. David Salle's "Tight as houses" is the most evident case of this where Harvey gives no indication that he has grasped the simple pun of the title and its clearly sexist content. Whose gaze is this painting painted from and for? Who could get the "joke"? The painting is treated with dead seriousness by Harvey, who cites Taylor (1987) on how it is a collage bringing together "incompatible source materials as an alternative to choosing between them" (Harvey, page 49). My own response, as someone who was potentially *in* that picture, and who saw it with completely different eyes, was: "here we go, another pretentious male artist who still thinks naked women are naughty."... The painting assumes a complicit male viewer. (44–5)

In a note to the paperback edition, at the end of that chapter, Harvey responds to critiques such as Massey's. He seems stung by the criticism. He writes:

> The illustrations used in this chapter have been criticized by some feminists

> of a postmodern persuasion. They were deliberately chosen because they allowed comparison across the supposed pre-modern, modern, and postmodern divides. The classical Titian nude is actively reworked in Manet's modernist Olympia … All of the illustrations make use of a woman's body to inscribe their particular message. The additional point I sought to make is that the subordination of women, one of many "troublesome contradictions" in bourgeois Enlightenment practices, can expect no particular relief by appeal to postmodernism. (65)

Harvey failed to recognize that using images that objectify women and empower the privileged position of the presumed male spectator require explicit response. He also chose to represent only the work of white men (though the creators of Citizen Watches are anonymous). Resistance required either comment or intentional juxtaposition with more liberatory images.

Michael Harris's *Colored Pictures: Race and Representation*

Michael Harris's *Colored Pictures: Race and Representation* makes up for Harvey's silence. He is alert to racial and sexual representation as he updates the genealogy of reclining nudes for the twentieth century. He also explicitly examines race as it is inflected in the genealogy. Harris observes that "even in black communities, black was a negative signifier." He points out that "blackness, unlike Jewishness or Irishness, is primarily visual." He elaborates on this visual and negating nature of blackness: "Racial discourses, although they are discourses of power, ultimately rely on the visual in the sense that the visible body must be used by those in power to represent nonvisual realities that differentiate insiders from outsiders" (2).

Toni Morrison's *Playing in the Dark* brilliantly illustrates Harris's insight about blackness as a negative signifier. She describes how, in conceptualizing ideas like freedom, nineteenth-century American whites needed the "unfree" black person to represent the "not-me." She writes: "It was not simply that this slave population had a distinctive color; it was that this color 'meant' something" (49). This leads Morrison to a discussion of the ending of *Huckleberry Finn*, in which Jim is not freed. She observes the "parasitical nature of white freedom."

Does Manet's *Olympia* betray this parasitical nature? A prostitute replaces Titian's *Venus*. In place of a white servant in Titian, Manet places an African woman behind the white woman, helping her. The white woman looks out of the painting toward the viewer; the African woman looks at the white woman.

Because of the way Manet brings race into his painting, Harris's commentary fills in the gap that Harvey's "no comment" methodology left unsaid. Harris carefully interprets the genealogy Harvey visually presented. What Harvey trusted his readers to do, Harris does not leave to chance—or interpretation. Moreover, he explicitly identifies the racial hegemony being inscribed. His chapter, "Jezebel, *Olympia*, and the Sexualized Woman," describes how "During the nineteenth century, the black female body in art had become a signifier of sexuality, among other things, as myths of black lasciviousness became entwined with other sexual ideas."

The black servant is not just a signifier of sexuality, but also for disease, an association that had been "part of the essentialist stereotyping of nonwhite women" (126). Further, "In the nineteenth century, women of color were associated with nature, uncontrolled passion, and promiscuity."

Let's note that *not* just in the nineteenth century have women of color been associated with nature, uncontrolled passion, and promiscuity, often by presenting them as wild, rather than domesticated animals. Annette Gordon-Reed reminds us that "The portrayal of black female sexuality as inherently degraded is a product of slavery and white supremacy, and it lives on as one of slavery's chief legacies and as one of white supremacy's continuing projects" (319).

In *Olympia*, Harris argues, "within the privileged space of the white male gaze is a layered black subject who is at once socially inferior to a naked prostitute, for whom she is a servant, and yet a sexual signifier and cipher; her mere presence is the equivalence of Olympia's nakedness" (126). She is the "not-me" of the painting; her imputed degraded sexuality suggesting a different kind of "not-free" status.

In his discussion of the female nude, Harris moves backward in time from Manet's *Olympia* to Titian's *Venus*. He finds three recurring aspects in this popular classical subject in Western art: evidence of patriarchal structures; the assumption of the universality of the white male perspective; the appropriation of female bodies (126).

Harris finds these three characteristics of Western art functioning in Titian's painting, and then, moving forward, exposes it in various artworks of the nineteenth century. He identifies compositional strategies that emphasize the visual availability of the woman being depicted, specifically a vertical line that thrusts downward to the vaginal area. Harris points out that this vertical line highlights the genital area of the nude woman. In fact, the composition of Titian's and Manet's paintings also includes a horizontal line made by the arm of the woman. That line, too, moves toward the pubic area of the white woman. Here is the place where

2010. Sistah Vegan: Food, Identity, Health, and Society: Black Female Vegans Speak, edited by A. Breeze Harper, published.

the vertical and the horizontal intersect. (It is reproduced with "Ursula Hamdress," too.)

As for the way Titian and Manet portray their subject's eyes, "The fact that each woman meets the viewer's gaze suggests that both women are complicit and compliant in the sexual arrangements" (129).

Harris then turns to the ways colonialism/imperialism inflected the image of the sexually consumable woman. He considers Jean Ingres's *Odalisque with a Slave,* and how it depicts "primitive sexuality" to be found outside Europe. Harris says: "All the painted harem nudes are available for visual consumption by the viewer—who is implicitly a European male" (130).

Linda Nochlin notes how paintings such as Ingres's "body forth two ideological assumptions about power: one about men's power over women; the other about white men's superiority to, hence justifiable control over, inferior, darker races" (Chadwick, 199).

Regarding the odalisques, what you see is what you get: both visual and literal sexual consumption in the service of and confirming the imperialist practices of the West. Harris writes: "All the nudes as paintings were available as artifacts for ownership and private consumption by male patrons, a fact that rehearsed or reiterated colonial and imperial adventure and its appropriation of land, resources, and people" (130).

Harris's genealogy then claims a place for Paul Gauguin's *Spirit of the Dead Watching.* Harris sees this as an inverted *Olympia*: "the black attendant and the reclining woman [now on her stomach] have become one" (131). The reclining woman is a pubescent girl. She is younger, more demure, less frank in her gaze, and fully submissive. The white bedclothes echo Titian's painting.

Pablo Picasso's *Les Demoiselles d'Avignon (*1907) is the next painting in Harris's genealogy. White prostitutes are depicted with African-themed masks, and though the white women are holding various poses, most of them standing, their poses suggest reclining nudes. By placing African mask-like faces on several prostitutes, Picasso merged "the primitivized white female and the imagined libidinous black woman into one body." Picasso's painting creates a black invisibility: "The mask forms insinuated a black presence yet subordinated it to the white females whose sexual consumption was linked to the colonial physical consumption of Africa" (131).

Harris's genealogy of the reclining nude ends with a photograph, "A South Sea Siesta in a Midwinter Concession," that was displayed at the California Midwinter International Exposition in 1894.[2] The non-Anglo woman is shown reclining on a mat—her breasts are uncovered, though her genital area is covered. "The photo shows the woman in a pose from a male-dominated discourse, but she gives no evidence of a willing compliance" (132). That was her power as a photographed subject: to refuse. The subjects of the paintings and the Citizen Watches ad, if they exercised such control in their gaze, could not control the representation of their gaze.

But, what if a photographed subject has no power to refuse? What if the photographed subject is an other-than-human animal either sedated or dead? Her eyes will be closed, yet there is no resistance in this act.

Harris helps us place the genealogy of the reclining nude within the context of oppressive attitudes regarding sex and race, and how these inter-sected. From Harris we can see how juxtaposition and superimposition function not only as artistic strategies, but as ways that complicate and confirm oppressive situations.

Nell Painter's *History of White People*

Intersectionality is always happening. It is happening with whiteness, too. It's just that whiteness, having been under-theorized for so long, often needs something else to put a spotlight on it. I will argue that when *Playboar* positioned a pink pig in the classic reclining nude pose, this became the "something else" that can turn a spotlight toward how whiteness is (and is not) functioning.

Nell Painter, in her *History of White People,* explores the ways in which whiteness came to signify power, prestige, and beauty. Though its staying power is remarkable (Roediger), whiteness was never monolithic in its granting of power and prestige. Whiteness, Painter reveals, also had a component of the "not-me" and the "not-free." She discerns two kinds of slavery in the anthropological work of eighteenth-century "science of race" scholars. Enslaved peoples whose bodies were forced to perform brute labor, including Africans and Tartars, were represented as ugly. But, luxury slaves also existed: they were "valued for sex and gendered as female" and became representations of human beauty. Painter describes how the term *odalisque* (and accompanying terms) carry with them "the aura of physical attractiveness, submission and sexual availability—in a word, femininity. She cannot be free, for her captive status and harem location lie at the core of her identity."

Painter examines the migration of the idea of "Causasian" beauty, and how it spread across the English Channel into Britain; the uncontested fact was that beauty resided in the enslaved. Georgian, Circassian, and Caucasian were all interchangeable names, so that in 1864 when P. T. Barnum asked his European agent "to find 'a beautiful Circassian girl' or girls" to exhibit in his

2010. VINE expands to begin to take in dairy industry survivors (including both former dairy cows and their cast-off sons), which they view as an explicitly feminist project.

New York Museum, Painter says, "In the American context, a notion of racial purity had clearly gotten mixed up with physical beauty." But when they arrived, Barnum's "Circassian slave girls" had the appearance of light-skinned Negroes as they all had "white skin and very frizzy hair." These white-skinned, frizzy-haired women offered a way to reconcile "conflicting American notions of beauty (that is, whiteness) and slavery (that is, Negro)" (51).

The science of race and the evolution of white slavery as a beauty ideal migrated into an attractive subject for male artists. As Painter describes, "By the nineteenth century, 'odalisques,' or white slave women, often appear young, naked, beautiful, and sexually available throughout European and American art" (43).

One cannot ignore Ingres if discussing odalisques, so like Harris, Painter turns to his paintings. Ingres's work "was a sort of soft pornography, a naked young woman fair game for fine art voyeurs." Painter notes that while the "odalisque still plays her role as the nude in art history," her part "in the scientific history of white race has largely been forgotten" (43).

Jean-Léon Gérôme's *Slave Market* also comes under review. In this painting, a young naked girl is being exhibited for sale. Her stance recalls (or prefigures) that of Picasso's prostitutes. Her pelvis tilts or rolls (suggesting submission as well as availability [see Adams 2004, 106]).

Painter notes the existence of slavery today, an issue that Nicolas Kristof and Sheryl WuDunn tackle in the first chapter ("Emancipating Twenty-First Century Slaves") of their book, *Half the Sky*. Their conservative estimate is that there are three million women and girls (and a small number of boys) who are enslaved in the sex trade (10).

Like Harvey and Harris, Painter finds a genealogy of images that carry forward into modernism. Two recent books, Edward Said's *Orientalism* and Anne McClintock's *Imperial Leather,* used white slave iconography on their book covers, though neither book dwelt on white slavery. Regarding this, Painter concludes, "Late twentieth-century American scholars seemed unable to escape Gérôme or confront slavery that was not quintessentially black" (56).

What else is a negative signifier beside blackness? Animality. Who else in contemporary society is enslaved besides women, girls, and an unknown number of boys? Other animals; the largest number being farmed animals. And so, whether we follow Harris, Harvey, or Painter, we arrive at "Ursula" and confront a slavery that is neither black nor human.

I created a chart to track the genealogy of the "Reclining Nude."

Genealogy of the "Reclining Nude" and cultural commentary on gender, race, species

	Harvey	Harris	Painter	Adams
Salle's *Tight as Houses*	✗ (photo with etchings)			
Titian's *Venus*	✗	✗		✗
Manet's *Olympia*	✗	✗		✗
Ingres's *Odalisque*		✗ with a slave (1840)	✗ *Grand* (1819) ✗ *Le Bain Turc* (1862)	✗
Powers's *The Greek Slave*			✗ (sculpture)	
Gérôme's *Slave Market*			✗	✗
Gauguin's *Spirit of the Dead*		✗		✗
Picasso's *Les Demoiselles*		✗		
South Sea Siesta		✗ (photo)		
Matisse's *Odalisque*			✗	
Rauschenberg's *Persimmon*	✗			
Citizen Watch	✗ (photo)			
Ursula/Taffy				✗ (photo)

All of Harvey's examples are of white women. Harris demonstrates the presentation of the reclining nude and what happens when one watches for the way racial and gender attitudes are inscribed together, so that African women represented animality. Painter considers a specific kind of presentation: that of white beauty as it was related to the enslavement of white women. I am looking for the way gender and race leap over the species line and become represented in a "reclining nude" that has a pig posed in a way similar to Titian's *Venus d'Urbino* and her successors. Harris notes "a fluidity between popular culture and fine art that gains momentum in the mid-nineteeth century and is taken for granted at the beginning of the twenty-first century" (11). *Playboar* fits the bill.

Harvey's, Harris's, and Painter's critiques are important, but the mistake of their critiques (though whether Harvey's "no comment" is a critique or

2010. Our Hen House, an online resource and multi-media magazine founded.

not is open to debate) is to believe that there is any sort of (human-based) closure to this genealogy.

Critical theory that investigates traditional Western depictions of women's beauty, specifically "the reclining nude," and follows these depictions into the late twentieth century has failed to recognize specifically how this tradition has leapt/fled/transcended human-centered notions to reinscribe retrograde and oppressive attitudes toward women and domesticated animals. My argument is that any of these critical theorists who think that tracing this genealogy can succeed when it is looking only at depictions of homo sapiens, especially female homo sapiens, has missed an interesting and important aspect of the genealogy. This aspect reveals how delving past the species line in representation of female "beauty," or sexualized female bodies, exposes the structuring of consumption of not just women, but domesticated animals. It is normalizing and naturalizing this consumption because it has fled the human without discarding representational aspects of race, sex, and class. Nonanthropocentric cultural theory will acknowledge how the sexualizing and feminizing of bodies intensifies oppressions on all sides of the species boundary.

The function of animalizing and racializing: That's why a pig

The way in which "Ursula" a pig is substituted for the woman reveals how overlapping absent referents that animalize, sexualize, racialize, and figure "youthfulness" interact.

In both Titian's and Manet's paintings we can find an animalizing function that exists parallel to the white woman at the center of the canvas: a little dog at Venus's feet in Titian, the African woman servant *and* a black cat in Manet.

When *Playboar* intervenes into this genealogy and places a female pig smack dab in the center of its staged photograph, the animalizing function has moved from margin to center. The animalizing and sexualizing functions which are separate in Titian's and Manet's paintings are united in one being.

With Harris and Painter providing foundational insights, we notice what we might not have noticed at first when we consider "Ursula": "Ursula" is marked as white. The white slave, the odalisque, available for sexual consumption has become the "white" enslaved female, available for literal consumption.

Once we notice how the pig's pink skin recalls white slavery, we realize why the pig's "racial" characteristics matter. Her whiteness is an anthropocentric anchor. As I discuss in *The Pornography of Meat*, if it were a "colored" pig (and after all, pigs can be many different colors), the non-dominant associations (gender, species, *and* race) would have been so great, there would be no anthropocentric hook. Because of the race hierarchy that still is inscribed so strongly in Western culture, a white pig was needed, so that the degradation being represented could be as strongly conveyed as possible (i.e., the whiteness associated with the pig, which normally would have provided a racial elevation, is contained/ overwhelmed by the female, animal, and enslavement associations). In addition, a "colored" pig would not have evoked the tradition of the *odalisque* and its figuration of whiteness. When we consider "Ursula Hamdress"—this popular culture manifestation of misogyny and objectification—we cannot ignore the racist figuring of white beauty that it is also drawing on while perverting it. The key here is that white beauty has a history tied to enslavement as well.

Harris's genealogy culminates in a photograph, as does mine: his, the "siesta-taker"; mine, "Ursula." Harris writes that the photograph suggests "a willingness by the photographer to contribute to the existing tradition of artistic nudes. Though this tradition was largely absent from American art, the nudity of a primitive nonwhite woman was more acceptable than that of a white woman" (132).

Why? Why was it more acceptable to photograph a primitive nonwhite woman? Harris suggests it is because it "offers a stage to play out white moral superiority." Visual consumption also provided distance from—yet enjoyment of—this dangerous, erotic-laden woman.

So, too, with a pig. The moral superiority is that of the human male; the visual consumption is of whiteness and (a farcical) "beauty"; the photograph offers the same distance and yet enjoyment of this very familiar, but now erotic-laden pig. And so the tradition moves whiteness to the nonhuman. With "Ursula," the photographer and all those who contributed to the creation and execution of the photograph express cynicism, while achieving detachment *and* enjoyment.

"Ursula Hamdress" is probably one of the founding images of anthropornography. Anthropornography is a neologism coined by Amie Hamlin and introduced in *The Pornography of Meat* to identify the specific sexualizing and feminizing of animals, especially domesticated animals consumed as food. Animals in bondage, particularly farmed animals, are shown "free," free in the way that "beautiful" women have been depicted as "free"—posed as sexually available as though their only desire is for the viewer to want their bodies. (Especially when such freedom was a lie.) They become the "not-free free." *Playboar* puts a face on meat eating that encapsulates a heinous, deeply offensive history of enslavement, misogyny, and racism. Anthropornography opens another avenue for these freighted meanings and images to be disseminated in and through popular culture.

Harris concludes his examination of the reclining nude as presented by artists representing the dominant culture with a discussion of voyeurism. He draws on the work of David Lubin who argues that "gazing at women voyeuristically is a means by which men may experience, reexperience or experience in fantasy their virility and all the potency and social worth that implies. Voyeurism by any definition, suggests detachment, estrangement, viewing from a distance" (134).

Harris extends Lubin's insight: "Voyeuristic engagement with the black/primitive woman safely separates the viewer from her dangers and reinforces his position within acceptable boundaries. He is white and gazes at the spectacle and danger of nonwhiteness, and he has the option of making real but disaffected forays into this realm as an exercise of his male prerogatives and power. Using the nonwhite female body as a spectacle ... offers a stage to play out white moral superiority because the exotic woman is a sign of the wanton sexual danger that white society has mastered" (134).

Which brings us back to the National Pork Producers' meeting in Des Moines and the excited buzz around looking at "Ursula."

What was to be gained by the voyeuristic experience of encountering "Ursula"?

Several things:

The cues that the largely male attendees were encountering were of a pornographed pig, so that in public they could do what, generally, with pornography, one did in private.

They would recognize that this pig had not endured life in one of their factory farms. So, for them the depiction of "Ursula" feeds on humor of the dominant culture about the one who "escapes." They would have possessed the cynical knowledge about how few pigs actually would have such unbruised skin as she did.

With "Ursula," there is the voyeurism not of the nonwhite (human) female body, but of the (ostensibly) white (nonhuman) female body.

"Ursula" would have been the real "pork producer"—her reproductive labor the necessary slavery for future piglets.

The human male exceptionalism that benefits from the positioning of Ursula/Taffy was further reinforced because *Playboar* was for sale at a public event (the National Pork Producers Council), and the participants could take the publication home and introduce it to their private space. The privileged male consumer knew that he was never going to be the enslaved consumed.

Harvey suggests that women's objectification is not resolved by postmodernism. The past 40 years have also been a time in which meat eating has been regressively associated with masculinity.

The postmodern representation might or might not resist complicity with the three points of Harris's recognized oppressive framework (patriarchal attitudes, white male as normative viewer, appropriation of female body), though Harvey clearly indicates that he did not believe it did resist this, nor was the postmodern intervention going to provide any particular relief. However, anthropornography does not resist it; it is not only complicit in this oppressive approach towards representing women, it simultaneously hides and celebrates its complicity, simultaneously makes fun of itself and never truly resists the figuration—consumption, it seems to say, is consumption and the "carnivorous virility" (Derrida) that constitutes the Western subject is okay by them.

This is the status quo not just reinscribed, but extended, compelled to sink, compelled to register the lowering that "carnivorous virility" is causing, this doubled interactive lowering, and "carnivorousness" is at the heart of it. The message made explicit: "This being is consumable." Just who *this* being is is fudged slightly, fudged to delight the virile carnivorous viewer.

Perhaps no area of representation intersects race, sex, and species as much as barbecue images. In *Making Whiteness,* Grace Elizabeth Hale argues that the "New South" of the early twentieth century constructed whiteness (and its enforcer Jim Crow laws) as an identity in response to the success of the Black middle class. Her work, like that of Harris and Painter, is concerned with racial making. Barbecues that use images of full-bodied white female sexual beings are a strange legacy of this constructed whiteness. With the images that advertise barbecues, what you see is what you get—visual and literal consumption of the full-bodied female body. They are "Ursula's" sisters and they share her fate.

Massey—who tackled Harvey's lack of critical consciousness—observes: "It is now a well-established argument from feminists but not only from feminists, that modernism both privileged vision over the other senses and established a way of seeing from the point of view of an authoritative, privileged, and male, position" (45). Massey continues: "The privileging of vision impoverishes us through deprivation of other forms of sensory perception." She then quotes Irigaray: "In our culture, the predominance of the look over smell, taste, touch, hearing, has brought about an impoverishment of bodily relations … the moment the look dominates, the body loses its materiality" (46).

Massey suggests something more is going on: "more important from the point of view of the argument here, the reasons for the privileging of vision is precisely its supposed detachment."

2011. A vegan-feminist Hunt Sab group formed to challenge traditional patriarchal rules of the countryside and related misogynist attitudes. They decide contextually when to display "male tropes" (bad language, extreme aggression), when to be silent, when to wear masks, when to wear makeup, when to smear menstrual blood on their faces. They work from the premise that they will be experienced as the wrong sex in the wrong place with the wrong ideas about animals.

This is what the posing and photographing of Ursula finally achieves—detachment from a body for whom materiality was everything but emptied of all meaning, a body whose role was to grow bodies for consumption and then be consumed itself.

Harris too, recognizes the implications of his genealogy for the voyeurism of the viewer: the act of looking becomes "the equivalence of sexual action; he is able to give the nude woman an ocular caress" (129).

After those pork producers looked at and laughed about "Ursula," after the ocular caress, they went home and returned to pigs whom they could touch, artificially impregnate, kill, and consume.

Acknowledgments

This essay was first given as a keynote address in 2010 at the Animals and Animality Graduate Student Conference at Queen's University in Kingston, Ontario. I thank the students at Queen's who invited me, and the participants with whom fascinating conversations occurred. I also thank Jim Mason for conversations with me in February 2012 in which we retraced our encounters with the image of Ursula Hamdress.

Notes

1   This essay continues my reflection on race and critical animal theory. Earlier essays can be found in Adams 1994, 2004, 2007, and 2012.
2   Harris believes the photograph was shown; I wonder if the woman wasn't on display there. As this event was an echo of and re-presentation of some of the exhibits of the Chicago World's Columbian Exposition, it would seem that they would have arranged to have people from non-dominant cultures on display, as they did in Chicago. Thus, it could be a photograph of a woman in the exhibit who was at the California Exposition.

Toward New EcoMasculinities, EcoGenders, and EcoSexualities

Greta Gaard

13

Are there masculinities that could be consistent with ecofeminist praxis? From years of organizing through the "chain of radical equivalences" among social movement actors, advocated by Ernesto Laclau and Chantal Mouffe (1985) as crucial to the formation of a radically democratic social movement, eco-justice activists and scholars have learned the value of deconstructing the role of the Dominant Master Self, and providing a location for even those constructed as dominant (whether via race, gender, class, sexuality, or nationality) to embrace a radically ecological vision and stand with—rather than on top of—the earth's oppressed majorities. For any egalitarian socioeconomic and eco-political transformation, such as that advocated by ecofeminism to be possible, both individuals and institutions need to shift away from overvaluing exclusively white, male, and masculinized attributes and behaviors, jobs, environments, economic practices, laws and political practices, in order to recognize and enact eco-political sustainability and ecological genders. Yet, while ecofeminist theory and praxis continues to be articulated by scholar-activists of diverse biological identities, genders, and sexualities, these scholars have not theorized the intersections between their embodiments and their ecofeminist praxis. In thinking about futures that

2012. To counter forced evictions, Niagara Falls-based Marineland Animal Defense provides anti-foreclosure work on property that was purchased by Marineland's owner.

don't center heteromasculinity and creating a more inclusive and descriptive ecofeminism—one that provides strategies and locations for co-creatively revisioning, educating, and mobilizing those who reject anti-ecological gender constructions—it is useful to start thinking about eco-masculinity.

Hegemonic masculinity requires an ecofeminist rethinking because humans (from industrial capitalists to ecofeminists and environmentalists) are gendered, sexual beings, and gender is crucial to many peoples' erotic expressions. As a scholar and activist, I am invested in the exploration of eco-masculinities because my own queer animal femme eco-sexuality prompts me to do so. In my own ecofeminism, gender and eroticism are entangled with my love of this earth. I want words for the butch resonance of rhyolite under my fingers when I am rock-climbing. I want language for the erotic attraction arising between my homecoming presence and feline greetings that approach and recede to rub against doorways and chair legs, eyeing me all the while. I want theory for the desire I inhale from a long-limbed lover who smells like trees. And I know I'm not alone in this eco-erotic bricolage of gender, species, nature. From Virginia Woolf's transgendered protagonist in *Orlando* (Woolf 1928) to Jeanette Winterson's unnamed and ungendered narrator in *Written on the Body* (Winterson 1992), queer feminist writers are envisioning eco-masculinities unbounded by sexual biologies, encompassing diverse sexualities and enacting diverse sexual practices.[1]

As Annie Sprinkle writes in *Bi Any Other Name*, "I started out as a regular heterosexual woman. Then I became bisexual. Now I am beyond bisexual—meaning I am sexual with more than just human beings. I literally make love with things like waterfalls, winds, rivers, trees, plants, mud, buildings, sidewalks, invisible things, spirits ..." (Sprinkle 1991, 103).[2] From Sprinkle's position as a former sex worker and porn star, and now self-described "eco-sexual," to Terry Tempest Williams's public position as Mormon heterosexual wife, the expression of eco-erotics and eco-genders suggests a wealth of information for considering eco-masculinities. In literature, Terry Tempest Williams's *Desert Quartet* (1995) describes eco-erotic encounters between a human hiker and the four elements. In the slot canyons of Utah's Cedar Mesa, the narrator's palms "search for a pulse in the rocks," while her body finds "places my hips can barely fit through" until "the silence that lives in these sacred hallways presses against me. I relax. I surrender. I close my eyes. The arousal of my breath rises in me like music, like love, as the possessive muscles between my legs tighten and release. I come to the rock in a moment of stillness, giving and receiving, where there is no partition between my body and the body of earth" (Tempest Williams 1995, 8–10). Hiking along a creek in the Grand Canyon, "only an hour or so past dawn," the narrator decides to take off her "skin of clothes" and leave them on the bank, lying down on her back and floating in water: "Only my face is exposed like an apparition over ripples. Playing with water. Do I dare? My legs open. The rushing water turns my body and touches me with a fast finger that does not tire. I receive without apology. Time. Nothing to rush, only to feel. I feel time in me. It is endless pleasure in the current" (23–4). Are the rock and the water gendered in these eco-erotic encounters? Or does the eco-erotic include and transcend gender?

Perhaps it is (past) time to envision alternative genders—and particularly eco-masculinities—from an ecofeminist perspective. What would it mean to redefine, or reconceive, an ecological masculinity?

Toward an ecological masculinity

Many ecofeminist philosophers, men's movement writers, animal studies and cultural studies scholars offer diverse yet mutually reinforcing critiques of Euro-Western cultural constructions of masculinity as predicated on themes of maturity-as-separation, with male self-identity and self-esteem based on dominance, conquest, workplace achievement, economic accumulation, elite consumption patterns and behaviors, physical strength, sexual prowess, animal "meat" hunting and/or eating, and competitiveness. These constructions developed in opposition to a complementary and distorted role for women: white hetero-human-femininity (Adams 1990; Buerkle 2009; Cuomo 1992; Davion 1994; Plumwood 1993; Schwalbe 2012). Recent studies of hegemonic masculinity as portrayed in men's lifestyle magazines confirm its pervasive representation via discourses of appearances (strength and size), affects (work ethic and emotional strength), sexualities (homosexual vs. heterosexual), behaviors (violent and assertive), occupations (valuing career over family and housework) and dominations (subordination of women and children) (Ricciardelli, Clow, and White 2010, 64–5). These representations varied far less than researchers expected among straight and gay-oriented men's magazines, and reaffirm the continuing force of hegemonic masculinity across sexualities and nationalities.

A term that was named "word of the year" in 2003 (Danford 2004), *metrosexuality* was articulated via the mass media television show, "Queer Eye for the Straight Guy," and *The Metrosexual Guide to Style* by Michael Flocker (Flocker 2003); both productions promoted "gay" advice for heterosexual men, emphasizing "self-presentation, appearance, and grooming" (Ricciardelli, Clow, and White 2010, 65). Beneath metrosexuality's "softened"

2012. A radical queer liberation group, Bash Back, responds to the problem of captive Orcas by recognizing solidarity between queer liberation struggles and the resistance of other animals and that their own liberation is directly tied to that of those suffering in our marine parks, zoos and beyond.

masculinity, scholars found the same hegemonic masculinity, influenced and intensified by consumerism, youth-obsession, and an emphasis on appearance, mandates usually enforced for femininity, and by association (Pharr 1988), for gay males as well: "whenever hegemonic masculinity is challenged, a new hegemonic form emerges," and thus "hegemonic masculinity actually becomes more powerful because of its ability to adapt and to resist change" (Ricciardelli, Clow, and White 65).

From bikini waxing to collagen injections and shopping (Frick 2004), metrosexuals were soon called back to hegemonic hetero-masculinity through beef consumption. Discussing Burger King's commercial "Manthem" as a textual narrative of gender, C. Wesley Buerkle (2009) reaffirms arguments made by Carol Adams (1990) that the very act of eating is associated with masculinity, and meat eating is an act of masculine self-affirmation.[3] Through advertising images and commercials, fast-food franchises such as Burger King and Hardee's portray hamburger consumption as enacting men's symbolic return to their supposed essence: personal and relational independence, nonfemininity, and virile heterosexuality. Continuing research reaffirms and uncovers further evidence to support Marti Kheel's critique of hegemonic masculinity's anti-ecological foundations, articulated in the ways it "idealizes transcending the [female-imaged] biological realm, as represented by other-than-human animals and affiliative ties" and "subordinate[s] empathy and care for individual beings to a larger cognitive perspective or 'whole'" (Kheel 2008, 3). The unstated fact that the more-than-likely "spent" dairy cows and other juvenile cattle slaughtered for fast-food beef hamburgers served at Burger King and Hardee's contribute exponentially to the accelerated rate of global warming (FAO 2006) underscores the anti-ecological impact of beef-eating hegemonic masculinity.

But masculinity has not always been defined in opposition to ecology. Although Lynn White's (1967) critique of Christianity's anthropocentric dominion over nature is probably the first and best-known, feminist and ecofeminist theologians such as Rosemary Radford Ruether (1983, 1992), Carol Christ (1997, 1979), Charlene Spretnak (1982), and Elizabeth Dodson Gray (1979) advanced beyond White's, offering significant critiques of monotheistic, patriarchal religions that worship a sky god and remove spirituality and the sacred from the earth, placing Hell beneath our feet and Heaven in the sky, deifying men, and valuing men's associated attributes over the values, attributes, and bodies of women, children, non-human animals, and the rest of nature. But prior to patriarchal, monotheistic religions, history and archeology show a different value was placed on women, nature, fertility, and the cycles of the earth. Following feminist theologians, men's movement scholars interested in mythology and archetypes make an important distinction between "sky god archetypes [who are] often warlike, either youthful invincible heroes, or older dominant males who rule in the name of an all-powerful [and often wrathful] sky god" and who serve to define masculinity as "a journey of ascension," and contrast them with earth gods: "archetypal images that relate masculinity to the earth" and offer "a different journey, one of descent, a 'going down' into, initially for many men, grief" (Finn 1998). In Arizona and New Mexico, the earth god is Kokopelli, the hump-backed flute player, a 3,000-year-old Hopi symbol of fertility, replenishment, music, dance, and mischief. In Europe, it is the Green Man, pictured as a male head disgorging vegetation from his mouth, ears, eyes; often associated with serpents or dragons, the Great Goddess, and the sacred tree, the Green Man dates back to Celtic art before Roman conquest, and to the work of Roman sculptors in the first century CE, and includes manifestations in figures such as Osiris, Dionysus, Cernunnos, and Okeanus (Anderson 1990).

But as many feminist spirituality groups have discovered, most people can't jump backwards in history, and attempts to re-enter and revive ancient traditions can seem not only ill-fitting, but also fail to provide maps and solutions for contemporary eco-social problems. Their value, however, lies in the fact that their presence proves there have been ecological, life-giving, and nurturing attributes associated with masculinity; thus, alternatives to hegemonic, anti-ecological masculinism may again be possible. Reconstructing ecological masculinities in 2012 and beyond, however, will require influence and insight from the last century of eco-justice movements, philosophies, and activisms.

Certainly, there have been significant silences that need to be addressed. As Mark Allister explains in *Eco-Man*, "gender studies in ecocriticism have been dominated by attention to feminism, [and] men's studies has been blind in seeing nature," citing the most important anthology in the field, Michael Kimmel and Michael Messner's *Men's Lives,* for supporting evidence (Allister 2004, 8–9). But despite its provocative subtitle, and its expressed intention "to serve as a companion to ecofeminism," *Eco-Man: New Perspectives on Masculinity and Nature* offers "no consistent underpinning" for its contents, and "no general deconstruction … of masculinity" (Allister 2004, 8). Similarly, in the premier volume on *Queer Ecologies* (Mortimer-Sandilands and Erickson 2010), while ample focus is given to human queer identities and other species' queer sexual practices, no attention is paid to the practices and organizations inspired by vegan lesbians, the presence and meaning of numerous websites and listservs for queer vegetarians, or the argument that vegan sexuality challenges heteronormative masculinity (Potts and Parry 2010); moreover, discussions of gender are relegated to

2012. *Hypatia: A Journal of Feminist Philosophy* publishes a special issue on "Animal Others."

a footnote summarizing Judith Butler's (1997) description of masculinity and femininity as "precarious achievements that are socially and psychically produced, in the context of a prohibition against homosexuality, through the *compulsory loss* of homosexual attachments," a loss that is "essentially melancholy in character" (Mortimer-Sandilands and Erickson 2010, 356n. 10). In sum, neither ecocriticism, nor men's studies, nor queer ecologies, nor (to date) ecofeminism has offered a theoretically sophisticated foray into the potentials for eco-masculinities.

Perhaps this omission is rooted in second-wave feminism's rejection of gender roles as universally oppressive. Radical texts of second-wave feminism such as June Singer's *Androgyny* (1977), which explores diverse religious and philosophical traditions from Plato's *Symposium* to the Book of Genesis, and from Jewish mysticism in the Kabbalah to Hindu practices of Tantra, concludes that both masculine and feminine traits are part of a whole and healthy psyche (meaning both soul and mind), and our job as self-actualizing humans involves "transcending" gender and "simply flowing between the opposites" (Singer 1977, 332). Yet such thousand-year-old constructions of gendered identity perpetuate the notion of dualized and polarized gender characteristics, advancing an essentialism that ecofeminists later rejected as limiting to theory-building and inclusivity (Davion 1994; Cuomo 1992). Instead of perpetuating the heterosexually distorted binary gender roles of masculine and feminine through an ideal of androgyny, or pretending that gender can be erased by eschewing all gendered cultural practices (from shaving and make-up to competition and weight-lifting), feminist eco-masculinity theorists need to reconceive gender—because we can't dismiss it. As a primary portal to the erotic, gender is more engaging when multiply expressed and freely crafted into diverse expressions. Moreover, exploring gender quickly leads to exploring sexualities, and opening the possibilities for not just eco-masculinities but eco-genders, eco-sexualities, and the eco-erotic.

As Judith Halberstam's groundbreaking volume on *Female Masculinity* (1998) describes, there is a long history of women-born-women with variously expressed masculine gender identities, from nineteenth-century tribades and female husbands to twentieth-century inverts and butches, transgender butches and drag kings. But the conjunction of *ecological feminist* politics and practices with these female masculinities has not been fully theorized, and the conjunctions range widely: while some articulations of female masculinity—most notably, some drag kings and transmen—have perpetuated oppressive manifestations of masculinity via sexism, exercising male privilege, and objectifying women, other articulations of transgender (and feminist) masculinities, exemplified in the annual Cascadia Trans and

Womyn's Action Camp, engage in everything from "ecosexual hikes" to "climb line rigging," "coalitions against coal," "nonviolent action training," "racism through an intersectional framework," and "self-care for cyclists" (Trans and Womyn's Action Camp 2012). Halberstam's research addresses the embodiment of female masculinities as performance and identity, providing groundwork for further attention to intellectual forms of female masculinities (i.e. theory-building, interrupting/contesting the corporate media), to the eco-political relation between butch identities and veganism, for example, or to climate justice and the material realities of economically marginalized women, people of color, queers, and non-human animals. As even Halberstam writes in her concluding chapter, "I do not believe that we are moving steadily toward a genderless society or even that this is a utopia to be desired" (272). Instead, theorizing the ecological articulations of a diversity of genders and sexualities may be a more strategic way to explore material dimensions of animal and ecological health. Although lesbian femme and hetero-feminine genders also have ecological intersections, I am drawn to exploring eco-masculinity because masculine gender identity has been constructed as so very *anti*-ecological, and thus its interrogation and transformation seem especially crucial. Moreover, the tools for this exploration are close at hand.

From ecofeminist theory, "boundary conditions" for eco-masculinities can be adapted to offer preliminary groundwork. For example, Karen Warren's eight boundary conditions of a feminist ethic (Warren 1990), as applied to eco-masculinity, might read: (1) not promoting any of the "isms" of social domination; (2) locating ethics contextually; (3) centralizing the diversity of women's voices; and (4) reconceiving ethical theory as theory-in-process which changes over time. Like a feminist ethic which is contextualist, structurally pluralistic, and in-process, an eco-masculinity would also strive to be (5) responsive to the experiences and perspectives of oppressed persons of all genders, races, nations, and sexualities; (6) it would not attempt to provide an objective viewpoint, knowing that centralizing the oppressed provides a better bias. As with feminist ethics, an eco-masculinity would (7) provide a central place for values typically misrepresented in traditional ethics (care, love, friendship, appropriate trust), and most significantly, (8) reconceive what it means to be human, "since it rejects as either meaningless or currently untenable *any gender-free or gender-neutral description of humans, ethics, and ethical decision making*" (Warren 1990, 141, italics mine). By rejecting abstract individualism, feminist eco-masculinities would recognize that all human identities and moral conduct are best understood "in terms of networks or webs of historical and concrete relationships" (141). Building on Warren's theory,

2012. Ahimsa House in Atlanta provides emergency pet safehousing, veterinary care, pet-related safety planning, legal advocacy, a 24-hour crisis line, outreach programs, and other services to help the human and animal victims across Georgia reach safety together.

an ecological masculinity would have to be explored through cross-cultural and multicultural perspectives to protect against privileging any specific race, region, or ethnicity. Patriarchy has shaped most contemporary industrial capitalist cultures, so eco-masculinities would need to recognize and resist the identity-shaping economic structures of industrial capitalism, its inherent rewards based on hierarchies of race/class/gender/age/species/sex/sexuality, and its implicit demands for ceaseless work, production, competition, and achievement. With ecofeminist values at heart, eco-masculinities would develop beyond merely rejecting the bifurcation of heterogendered traits, values, and behaviors: eco-masculinity/ies would enact a diversity of ecological behaviors that celebrate and sustain biodiversity and ecological justice, interspecies community, eco-eroticisms, ecological economics, playfulness, and direct action resistance to corporate capitalist eco-devastations. Already, developments are underway.

To date, Paul Pulé (2007, 2009) has been foremost in developing an "ecological masculinism" that replaces an "ethic of daring" (based on dominant male values such as rationality, reductionism, power and control, confidence, conceit, selfishness, competitiveness, virility) with an "ethic of caring" for self, society, and environment (with associated values of love, friendship, trust, compassion, consideration, reciprocity, and cooperation with human and more-than-human life). Optimistically, Pulé identifies eight key conceptual frameworks across the political spectrum, along with seven "liberatory ideals in sympathy with Leftist politics" that he believes support "a shift away from hegemonic masculinities and towards a long-term ecological sustainability"; he proposes an eco-masculinism that "may crucially contribute to this shift" (Pulé 2007).[4] While Pulé's work offers a foray into this discussion, he omits Plumwood (1993), Warren (1994, 1997, 2000), Salleh (1984, 1997), and many other ecofeminist critiques of several of his listed key conceptual frameworks and liberatory ideals—critiques that prove many of these conceptual frameworks to be inherently unsuited to even a *feminist* revisioning of ecologically oriented gender. Moreover, apart from a footnote, Pulé does not consider the strong influences of race, class, sexuality, and culture in constructing masculinities.

In advancing a truly ecological and feminist masculinity, the heterosexism implicit in hegemonic constructions of masculinity would need to be resisted, drawing on insights and questions from the new queer ecologies (Gaard 1997; Mortimer-Sandilands and Erickson 2010). What would an eco-trans-masculinity look like? Are all Lesbian Rangers eco-butches, or are there eco-femmes flashing lesbian masculinities in the Parks Service as well?[5] Could we imagine eco-fags, radical faeries who dance and flirt and organize for eco-sexual justice?

Indeed we can. In the 1960s and 1970s, gay liberation activists seeking ways to articulate the intersections of gay sexuality, spirituality, eco-anarchist politics, and genderfuck created the Radical Faeries. Describing themselves as "a network of faggot farmers, workers, artists, drag queens, political activists, witches, magickians, rural and urban dwellers who see gays and lesbians as a distinct and separate people, with our own culture, ways of being/becoming, and spirituality," Radical Faeries believe in "the sacredness of nature and the earth [and] honor the interconnectedness of spirit, sex, politics and culture" (Cain and Rose). They include legendary queer visionaries such as Harry Hay, Will Roscoe, and Mitch Walker, with their history and vision articulated through Arthur Evans' *Witchcraft and the Gay Counterculture* (Evans 1978). Within their regionally placed communities and annual gatherings, Radical Faeries celebrated an earth-based spirituality that honored sexuality and began the applied work of articulating contemporary, non-hegemonic eco-masculinities. Describing themselves as "not-men," "sissies," and "faeries," the Radical Faeries' "manifesto" offered only a short statement about feminism: "As faeries we are very interested in what our sisters have to say. The feminist movement is a beautiful expansion of consciousness. As faeries we enjoy participating in its growth" (Cain and Rose). Unfortunately, the faeries continue to describe the earth as female, a gendering that ecofeminists have shown tends to perpetuate Eurocentric gender stereotypes (i.e. earth as nurturing mother who will clean up men's toxic wastes, as a bad and unruly broad who brings hurricanes and other "bad" weather, or a virgin to be ravished/colonized, etc.) and does not improve real material conditions for women or nature (Gaard 1993). Nonetheless, the Radical Faerie movement launched interrogations of queer eco-masculinities that have been advanced over the past four decades.

In "Wigstock" (1995), a documentary covering the annual drag festival in New York City, emcee Lady Bunny says, "I think Mother Nature must be a Drag Queen," articulating the nexus of eco-masculinity and genderfuck that drag queens are well positioned to provide. Consider the "radioactive" drag queen Nuclia Waste (Krupar 2012), whose flamboyant performances draw attention to the clean-up efforts at former plutonium production facility Rocky Flats, Colorado, that have converted this location into a wildlife refuge. As a triple-breasted and sparkly-bearded drag queen with glowing green hair, Nuclia Waste makes visible "the porosity of body and environment and the ways humans and nonhumans have been irrevocably altered by nuclear projects"(Krupar 2012, 315). Her digital performances reintegrate toxic waste, mutant sexualities, and popular culture, "queering the nuclear family" and encouraging viewers to consider "the entire US as a nuclear landscape" and the pervasive "presence of nuclear waste in everyday

2012. Conference celebrating Marti Kheel's vision for an inclusive ecofeminist theory and practice held at Wesleyan University.

life" (316). Mixing stereotypically male and female signifiers, Nuclia's drag performs an irreverent critique of nature/culture and waste/human binarisms, insisting on the impossibility of purity and queering humans as "boundary-creatures, neither fully natural nor fully civilized" (317). Yet her work could do more to address the real material conditions for "wildlife"—i.e. the more-than-human animals reintroduced to clean up appearances at this nuclear waste site. In sum, Nuclia's performances offer an embodied ecological politics that is crucial to an ecofeminist reconsideration of eco-genders and eco-sexualities—and interspecies ecologies need to be central in such reconsiderations.

Introducing the term "ecogender," Banerjee and Bell (2007) argue that "women and men have been interacting with the environment for ages, *qua women and men*, without consciously attempting to do so" (Banerjee and Bell 2007, 3). Although their research accepts sexual and gender dualisms, they offer an environmental social science critique of ecofeminism that is helpful to this project of constructing eco-masculin-ities: "Merchant's [1980] view of precapitalist society passes easily over the brutality of feudal hierarchies," they observe, and "Plumwood [1993] does not identify the logic of domination outside of the West" even though patterns of dominating women, non-dominant men, children, more-than-human animals and nature can readily be found in non-Western societies. Moreover, "Mellor's [1992, 1997] vision of women as environmental mediators homogenizes women's experience and unnecessarily excludes men as potential mediators," and "Salleh [1984, 1997] does not confront the question of the commodification of men and male labor" (7–8). Far from articulating the anti-feminist complaint that "men are oppressed too!", Banerjee and Bell remind us that the elevation of a few elite men has been advanced at the expense of other less-dominant men, women, children, animals, and the environment. As Warren's boundary conditions suggest, liberatory theories that exclude or overlook the oppression of any subordinated group cannot hope to provide a holistic description of the logic and functioning of oppressive systems, or propose effective strat-egies for their transformation. Based on the understanding that "gender itself is a relational construction, and that therefore women's and men's embodied environmental experience cannot be understood in isolation" but must be historically and culturally situated, Banerjee and Bell propose an eco-gender study to explore "the dialogic character of the relationality of gender, society, and environment" which will uncover "the patterns of oppression that constrain these interactions" (14). Although their study omits consideration of sexualities and species relations, their articulation of eco-gender as an encompassing approach to bringing ecofeminist theory into the environmental social sciences directly addresses Kheel's critique of hegemonic masculinity.

Putting these diverse approaches together raises questions about the ecological implications of gender and sexuality alike. Approaching eco-gender from the perspective of bisexuality, Serena Anderlini-D'Onofrio argues that our "current erotophobic cultural climate" can be disrupted by bisexual practices that function "as a portal to a world without the homo-hetero divide," unleashing an erotophilia whose "transformative force" can power more loving and ecologically effective responses to climate instability and a variety of human health crises (Anderlini-D'Onofrio 2011, 179, 186, et passim). To move beyond the bedroom into the sociopolitical, or from the erotic to the eco-erotic, this erotophilia needs to be linked with an eco-anarcha-feminist political approach.[6]

Clearly, the humanist (or as Kheel would argue, the anthropocentric and more specifically *andro*centric) orientation of most culturally constructed masculinities must be interrogated. Disentangling biological sex, gender expression, gender role, sexual orientation, and sexual practices, as queer studies scholars like Anderlini-D'Onofrio and others suggest, can we describe (not define) diverse eco-sexualities that play fast and loose with gender while actively working for environmental, interspecies, and climate justice? How might a queer, interspecies consideration of gender guide our revisioning of human eco-masculinities and eco-sexualities?

Eco-masculinities, ecosexualities, and their contemporary expressions

For examples of eco-masculinities and the eco-erotic, we can look to the music and lyrics of "Nature Boy" eden ahbez, ecocritic Jim Tarter, and Saami-American artist Kurt Seaberg.[7] In 1947, eden ahbez (who always spelled his name in lower case) approached Nat King Cole's manager in Los Angeles, and handed him the music and lyrics for "Nature Boy," a song that quickly became famous. A disciple of Paramahansa Yogananda's silent meditation practices, ahbez lived a life of economic and ecological simplicity, wearing burlap pants and sandals, sleeping outdoors underneath the Hollywood sign, and eating a vegetarian diet. He later lived in community with other like-minded yogis in Laurel Canyon, and collaborated with jazz musician Herb Jeffries on his "Nature Boy Suite." Predating the hippie movement of the 1960s, ahbez performed bongo, flute, and poetry gigs at beat coffeehouses in the Los Angeles area. In 1960—the year I was born—ahbez recorded his only solo LP, *Eden's Island,* for Del-Fi Records. Growing up in a Los Angeles suburb in the 1960s, I listened to eden ahbez's album "Eden's Island" and played over and over again the songs "Full Moon" and "La Mar" for their ecological economics and eco-spirituality, expressed in lyrics reminiscent of Walt Whitman's poetry.[8] It wasn't until adulthood that I researched ahbez and found he had been living two miles from my childhood home, where his ecological, spiritual, and political-economic ethics preceded and made space for my own vision of ecofeminist ethics.

Another vibrant example of feminist eco-masculinity can be seen in the life of ecocritic Jim Tarter. Writing in *The Environmental Justice Reader,* Tarter (2002) describes his battle with Hodgkin's Disease, a cancer of the lymphatic system, and his sister's battle with ovarian cancer. Quitting his career-track job, Tarter moved in with his sister Karen and became her primary, live-in caretaker for the last six months of her life. Together, they read Sandra Steingraber's *Living Downstream: An Ecologist Looks at Cancer and the Environment* (1997), and pieced together their family's battle with cancer in conjunction with the toxic environment of their early childhood years along the Saginaw River in Michigan, with General Motors and Dow Chemical and cement factories nearby. Through his caregiving of Karen, and their readings of Steingraber, Tarter realized that cancer is a feminist environmental justice issue for the ways it affects women's bodies, with the most dangerous carcinogens stored in body fats, and the cancers attacking women's reproductive organs (breast, uterus, ovaries). After Karen's death, Tarter continued his teaching and ecocritical scholarship in Idaho with this new focus, committing his teaching to educating indigenous students and bringing feminist perspectives into his work.

Lithographer, carpenter, gardener, playful actor and eco-activist writer, Kurt Seaberg makes his home by the Mississippi River, in a duplex he shares with African American poet Louis Alemayhu. Seaberg has replaced his entire back yard and driveway with a sustainable garden, where he grows much of his own food, and has planted native grasses around the front and sides of the home. Songbirds, hummingbirds, and wasps receive equal welcome in his garden with nests and feeders tucked among lattices and eaves, while mice find forage in his compost bins. As a visionary response to climate change, Seaberg uses his bicycle for transportation, and participates in local activist groups such as Friends of the Mississippi River, Occupy Minneapolis, Tar Sands Action, and supports the Indigenous Environmental Network. A former men's group participant and Green Party supporter, Seaberg brings a vision of social and ecological justice into his work, his creative and performing artistry, and his strong community ties. His artist's statement describes this work:

One of the tasks of the artist, I feel, is to remind us where our strength and power lies—in beauty, community and a sense of place. Nature has always been a theme and source of inspiration in my work, in particular the spiritual qualities that I find there. My hope is that my art will evoke the same feelings that arise in me when I contemplate the mystery of being alive in a living world: humility, gratitude and a sense of wonder before what I believe is truly sacred.

Based on these diverse examples from ecofeminist and queer ecological theories, literature, and lived experiences, what values, traits, and behaviors might articulate the radical possibilities of eco-masculinities, eco-gender, and/or eco-sexualities?

As Schwalbe (2012) reminds us, reconstructing hegemonic masculinity requires crucial actions linking the individual and the institutional: as examples, he suggests nurturing new minds in children, minds not oriented to seeking satisfaction in status, power, and the domination of others, nor in submission or blind obedience; and working to "end the exploitative economic and political arrangements that are sustained by a continuing supply of expendable men" (Schwalbe 2012, 42). Without the need to dominate and control others—and with the creation and ongoing presence of cooperative economic and democratic enterprises—there will be "little need for the kind of manhood that has evolved under capitalism" (44). Writing in *Eco-Man*, Patrick D. Murphy suggests another key feature of eco-masculinity: noting that "men are credited with creating but are not expected to nurture what they create," he laments that "nurturing remains a concept rarely applied to men and an area of male practice inadequately studied, discussed, and promoted" (Murphy 2004, 196–7). Murphy's essay explores some of the ways that fathers can nurture children while learning "to relinquish the illusion of control" and engaging with the fathers' own emotions, in dialogue with their children (208). Nurturing ecological sustainability, nurturing human and more-than-human companions, nurturing an ecophilic eco-erotic, and nurturing interspecies, ecological justice: these are some of the projects of a feminist eco-masculinity.

Inviting explorations: Eco-erotophilic-anarcha-feminist masculinities

Ecofeminist scholarship, and the research of scholars and activists across the disciplines, suggests that capitalist heteromasculinity is fundamentally anti-ecological. Of major significance in Kheel's *Nature Ethics* is her insight that ***all environmental ethics are constructed through the lens of gender.*** If environmental ethicists and activists want to make more conscious choices about that lens, we'll need to envision diverse expressions of eco-genders— not just eco-masculinities but also eco-femme and eco-trans identities—as well as eco-sexualities. As with our cultures, our physical, erotic animal bodies are a location of knowledge to be explored.

Acknowledgments

Special thanks to Lori Gruen and Chris Cuomo for suggesting I consider female (eco) masculinity.

Notes

1   For an ecofeminist ecocritical discussion of such gender-bending literatures, see Sharon Ruston, *Ecocriticism and Women Writers* (Palgrave Macmillan, forthcoming).

2   I excerpted this portion of Sprinkle's sentence because she completes her list by including "beings from other planets, the earth, and yes, even animals," raising the significant question of consent, and how one would determine consent from another species. Consent is a non-negotiable premise for all radical sexualities and sexual behaviors. For more on Annie Sprinkle and Elizabeth Stephens' work on ecosexuality go to: sexecology.org

3   The commercial is now publicly available on YouTube, at http://www.youtube.com/watch?v=R3YHrf9fGrw

4   Pulé's eight conceptual frameworks range from Progressive Left to Conservative Right, and include Socialist, Gay/Queer, Profeminist, Black (African), Mythopoetic, Men's Rights, Morally Conservative, and Evangelical; his seven liberatory ideals include Feminist Sociobiology, Deep Ecology, Social Ecology, Ecopsychology, Gaia Theory, Inclusionality Theory, and General Systems Theory.

5   Bruce Erickson (2010) introduces the Lesbian Rangers: "Shawna Dempsey and Lorri Millan founded the Lesbian National Parks and Services in 1997 as a way of inserting a lesbian presence into the natural landscape. In full uniform, the performance artists interact with the public, and point out potential hazards to the flourishing of lesbian flora and fauna in natural settings, including sexism and the

naturalization of heterosexuality in human and nonhuman contexts" (Erickson 2010, 328n. 3). Described as a group of "eager beavers," the Lesbian Rangers maintain a website with rich resources at http://fingerinthedyke.ca/index.html (accessed October 16, 2012).

6   Chaia Heller (1999) describes five dimensions of the socio-erotic, humans' desire for sensuality, association, differentiation, development, and political opposition as part of an eco-anarcha-feminist eroto-politics.

7   Robert Bly's masculinist and anti-feminist work is purposely omitted as it does not advance ecofeminist politics.

8   For a full listing of ahbez's lyrics, it is well worth the time to look at http://plus1plus1plus.org/Resources/eden-ahbez-lyrics

References

Acampora, Ralph. 1995. "The Problematic Situation of Post-Humanism and the Task of Recreating a Symphysical Ethos." *Between the Species* 11 (1–2): 25–32.

—2006. *Corporal Compassion: Animal Ethics and Philosophy of Body.* Pittsburgh: University of Pittsburgh Press.

Adams, Carol. 1975. "The Oedible Complex: Feminism and Vegetarianism." In *The Lesbian Reader*, Gina Covina and Laurel Galana (eds). Oakland, CA: Amazon.

—1976. "Vegetarianism: The Inedible Complex." *Second Wave* 4.

—1990. *The Sexual Politics of Meat.* New York: Continuum. 2010. 20th anniversary edition. London: Bloomsbury.

—1991. "Ecofeminism and the Eating of Animals." *Hypatia: Special Issue on Ecological Feminism* 6 (1): 125–45.

—1993. "The Feminist Traffic in Animals." In *Ecofeminism: Women, Animals, Nature*, edited by Greta Gaard. Philadelphia: Temple University Press, 195–218.

—1994a. *Neither Man nor Beast—Feminism and the Defense of Animals.* New York: Continuum.

—1994b. "Bringing Peace Home: A Feminist Philosophical Perspective on the Abuse of Women, Children, and Pet Animals." In *Hypatia: A Journal of Feminist Philosophy* 9 (2).

—1994c. "On Beastliness and a Politics of Solidarity." In *Neither Man Nor Beast: Feminism and the Defense of Animals.* New York: Continuum, 71–84.

—1995a. "Woman-Battering and Harm to Animals." In *Animals and Women: Feminist Theoretical Explorations*, Carol J. Adams and Josephine Donovan (eds). Durham and London: Duke University Press.

—1995b. "Comment on George's 'Should Feminists be Vegetarians?'" *Signs: Journal of Women in Culture and Society* 21 (1): 221–5.

—2004. *The Pornography of Meat.* New York: Continuum.

—2007. "The War on Compassion." In *The Feminist Care Tradition in Animal Ethics*, Josephine Donovan and Carol J. Adams (eds). New York: Columbia University Press, 21–36.

—2011. "After MacKinnon: Sexual Inequality in the Animal Movement." In *Animal Liberation and Critical Theory,* John Sanbonmatsu and Renzo Llorente (eds). Lanham, MD: Rowman and Littlefield.

—2012. "What Came Before *The Sexual Politics of Meat.*" In *Species Matters: Humane Advocacy and Cultural Theory*, Marianne DeKoven and Michael Lundblad (eds). New York: Columbia.

Adams, Carol J. and Josephine Donovan (eds). 1995. *Animals and Women: Feminist Theoretical Explorations.* Durham and London: Duke University Press.

Adler, Julius and Wung-Wai Tso. 1974. "'Decision'-Making in Bacteria: Chemotactic Response of Escherichia coli to Conflicting Stimuli," *Science,* n.s., 184 (4143) (June 21): 1292–4.

ahbez, eden. 1960. *Eden's Island: The Music of an Enchanted Isle.* Hollywood, CA: Del-Fi Records LP.

Alaimo, Stacy. 2010. "Eluding Capture: The Science, Culture, and Pleasure of 'Queer' Animals." In *Queer Ecologies: Sex, Nature, Politics, Desire*, Catriona Mortimer-Sandilands and Bruce Erickson (eds). Bloomington, IN: Indiana University Press, 51–72.

Allister, Mark (ed.). 2004. *Eco-Man: New Perspectives on Masculinity and Nature.* Charlottesville: University of Virginia Press.

Anderlini-D'Onofrio, Serena. 2011. "Bisexuality, Gaia, Eros: Portals to the Arts of Loving." *Journal of Bisexuality* 11: 176–94.

Anderson, Michael, dir. 1976. *Logan's Run.* Metro Goldwyn Mayer/United Artists.

Anderson, Virginia. 2006. *Creatures of Empire: How Domestic Animals Transformed Early America.* Oxford: Oxford University Press.

Anderson, William. 1990. *Green Man: The Archetype of our Oneness with the Earth.* London: HarperCollins Publishers.

Andrews, David (ed.). 2001. *Michael Jordan, Inc.: Corporate Sport, Media Culture, and Late Modern America.* Albany, NY: State University of New York Press.

Avital, Eytan and Eva Jablonka. 2000. *Animal Traditions: Behavioural Inheritance in Evolution.* Port Chester, NY: Cambridge University Press.

Avramescu, Catalin. 2010. Interviewed by Justin E. H. Smith, in "The Raw and the Cooked." *Cabinet Magazine.* Excerpted as "An Examination of Cannibalism Is Bound to Induce a Species of Metaphysical Unease," in *Berfrois: Intellectual Jousting in the Republic of Letters.* http://www.berfrois.com/2010/11/examination-of-cannibalism (accessed March 7, 2013).

Bad Newz Kennels, Smithfield Virginia. August 28, 2008. *Report of Investigation, Special Agent-in-Charge for Investigations Brian Haaser.* USDA Office of Inspector General-Investigations, Northeast Region, Beltsville, Maryland.

Bagemihl, Bruce. 1999. *Biological Exuberance: Animal Homosexuality and Natural Diversity.* New York: St. Martin's Press.

Bailey, C. 2007. "We Are What We Eat: Feminist Vegetarianism and the Reproduction of Racial Identity." *Hypatia* 22 (1): 39–59.

Baker, Steve. 2000. *The Postmodern Animal.* London: Reaktion Books.

—2001. *Picturing the Beast: Animals, Identity, and Representation.* Champagne and Chicago: University of Illinois Press.

—2013. ARTIST|ANIMAL. Minneapolis: Minnesota University Press.

Banerjee, Damayanti and Michael Mayerfeld Bell. 2007. "Ecogender: Locating Gender in Environmental Social Science." *Society & Natural Resources* 20: 3–19.

Barad, Karen. 2007. *Meeting the Universe Half-Way: Quantum Physics and the Entanglement of Matter.* Durham, NC and London: Duke University Press.

Barnosky, Anthony D., Elizabeth A. Hadly, Jordi Bascompte, Eric L. Berlow, James H. Brown, Mikael Fortelius, Wayne M. Getz, et al. 2012. "Approaching a State Shift in Earth's Biosphere." *Nature* 486 (7401): 52–8.

Baur, Gene. 2008. *Farm Sanctuary: Changing Hearts and Minds about Animals and Food.* New York: Touchstone.

Behnke, Elizabeth A. 1999. "From Merleau-Ponty's Concept of Nature to an Interspecies Practice of Peace." In *Animal Others: On Ethics, Ontology and Animal Life*, Peter H. Steeves (ed.). Albany, NY: State University of New York Press, 93–116.

Beirne, Piers. 1997. "Rethinking Bestiality: Towards a Concept of Interspecies Sexual Assault." *Theoretical Criminology* 1 (3): 317–40.

—1998. "For a Nonspeciesist Criminology: Animal Abuse as an Object of Study." *Criminology* 37 (1): 117–48.

—2004. "From Animal Abuse to Interhuman Violence? A Critical Review of the Progression Thesis." *Society and Animals* 12 (1): 39–65.

Bekoff, Marc and Colin Allen. 1998. "Intentional Communication and Social Play: How and Why Animals Negotiate and Agree to Play." In *Animal Play*, Marc Bekoff and John A. Byers (eds). New York: Cambridge University Press, 97–114.

Bekoff, Marc and Jessica Pierce. 2009. *Wild Justice*. Chicago, IL: University of Chicago Press.

Benney, Norma. 1983. "All of One Flesh: The Rights of Animals." In *Reclaim the Earth: Women Speak out for Life on Earth,* Leonie Caldecott and Stephanie Leland (eds). London: The Women's Press, 141–50.

Bentham, Jeremy. 1948. *An Introduction to the Principles of Morals and Legislation*. New York: Hafner/Macmillan.

Benton, Ted. 1988. "Humanism=Speciesism. Marx on Humans and Animals." *Radical Philosophy* 50: 4–18.

—1993. *Natural Relations: Ecology, Animal Rights, and Social Justice*. London and New York: Verso.

Berger, John. 1972 (2008). *Ways of Seeing*. New York and London: Penguin Books.

—1980 (1991). *About Looking*. New York: Vintage/Random House.

Bergson, Henri. 1974. *The Creative Mind: An Introduction to Metaphysics*. Trans. Mabelle Andison. New York: Citadel.

—1998. *Creative Evolution*. Trans. Arthur Mitchell. New York: Dover.

Bering, Jesse. 2012. "The Rat That Laughed." *Scientific American* 306 (7): 74–7.

Bérubé, Michael. 2010. "Equality, Freedom, and/or Justice for All: A Response to Martha Nussbaum." In *Cognitive Disability and its Challenge to Moral Philosophy,* Eva Feder Kittay (ed.). Hoboken, NJ: John Wiley & Sons.

Birke, Lynda. 1994. *Feminism, Animals and Science: The Naming of the Shrew*. Buckingham and Philadelphia: Open University Press.

Bishop, Sharon. 1987. "Connections and Guilt." *Hypatia* 2 (1): 7–23.

Blackwood, Evelyn and Saskia E. Wieringa. 2007. "Globalization, Sexuality, and Silences: Women's Sexualities and Masculinities in an Asian Context." In *Women's Sexualities and Masculinities in a Globalizing Asia*, Saskia E. Wieringa, Evelyn Blackwood, and Abha Bhaiya (eds). New York: Palgrave Macmillan, 1–21.

Bonilla-Silva, Eduardo. 2009. *Racism Without Racists: Color-Blind Racism and the Persistence of Racial Inequality in America*. 3rd edn. Lanham, MD: Rowman & Littlefield.

Browne, Kathe, Jason Lim, and Gavin Brown (eds). 2007. *Geographies of Sexualities: Theory, Practices and Politics*. Brookfield, VT: Ashgate Publishing Group.

Budiansky, Stephen. 1999. *The Covenant of the Wild: Why Animals Chose Domestication*. New Haven, CT: Yale University Press.

Buerkle, C. Wesley. 2009. "Metrosexuality Can Stuff It: Beef Consumption as (Heteromasculine) Fortification." *Text and Performance Quarterly* 29 (1): 77–93.

Burke, Bill. n.d. "Once limited to the rural South, dogfighting sees a cultural shift." http://hamptonroads.com/print/283641.

Butler, Judith. 1997. *The Psychic Life of Power: Theories in Subjection*. Stanford, CA: Stanford University Press.

—2004. *Precarious Life*. New York: Verso.

—2009. *Frames of War*. New York: Verso.

Cain, Joey and Bradley Rose. "Who are the radical faeries?". http://eniac.yak.net/shaggy/faerieinf.html (accessed October 16, 2012).

Callicott, J. Baird. 1989. *In Defense of the Land Ethic: Essays in Environmental Philosophy* Albany, NY: State University of New York Press.

Carastathis, Anna. 2013. "Basements and Intersections." *Hypatia* 28: 698–715.

Casselton, Lorna A. 2002. "Mate Recognition in Fungi." *Heredity* 88 (2) (February): 142–7. doi:10.1038/sj.hdy.6800035.

Cavell, Stanley. 1998. "Companionable Thinking." *Philosophy and Animal Life*. New York: Columbia University Press.

Cavendish, Margaret. 1972. *Poems, and Fancies*. (1653). Yorkshire: Scolar Press.

Chadwick, Whitney. 1990 (1996). *Women, Art, and Society*. London: Thames and Hudson.

Chou, Wah-Shan. 2001. "Homosexuality and the Cultural Politics of Tongzhi in Chinese Societies." *Journal of Homosexuality* 40 (3–4): 27–46. doi:10.1300/J082v40n03_03.

Christ, Carol P. 1997. *Rebirth of the Goddess: Finding Meaning in Feminist Spirituality*. New York: Routledge.

Christ, Carol P. and Judith Plaskow (eds). 1979. *Womanspirit Rising: A Feminist Reader in Religion*. New York: HarperCollins.

Coakley, Jay. 1998. *Sport In Society*. 6th edn. New York: McGraw-Hill.

Coetzee, J. M., et al. 1999. *The Lives of Animals*, Amy Gutmann (ed.). Princeton, NJ: Princeton University Press.

Cole, Eve Browning and Susan Coultrap-McQuin (eds). 1992. *Explorations in Feminist Ethics: Theory and Practice*. Bloomington: Indiana University Press.

Collard, Andrée with Joyce Contrucci. 1988 (1989). *Rape of the Wild: Man's Violence Against Animals and the Earth*. London: The Women's Press. Bloomington: Indiana University Press.

Collins, Patricia Hill. 2005. *Black Sexual Politics: African Americans, Gender, and the New Racism*. New York: Routledge.

Corrigan, Robert W. 1965. *Comedy, Meaning and Form*. San Francisco, CA: Chandler Publishing.

Crenshaw, Kimberlé Williams. 1989. "Demarginalizing the intersection of race and sex: A black feminist critique of antidiscrimination doctrine, feminist theory and antiracist politics." *University of Chicago Legal Forum*: 139–67.

—1992. "Whose Story Is It Anyway? Feminist and Antiracist Appropriations of Anita Hill." In *Race-ing Justice, Engendering Power: Essays on Anita Hill,*

Clarence Thomas, and the Construction of Social Reality, Toni Morrison (ed.).
     New York: Pantheon Books.
Cronin, Katie, Edwin van Leeuwen, Innocent Chitalu Mulenga, and Mark Bodamer.
     2001. "Behavioral response of a chimpanzee mother toward her dead infant."
     American Journal of Primatology 73 (5): 415–21.
Cuomo, Chris. 1992. "Unravelling the Problems in Ecofeminism." *Environmental*
     Ethics 15 (4): 351–63.
Cuomo, Chris and Lori Gruen. 1998. "On Puppies and Pussies." In *Daring to Be*
     Good, Ami Bar On and Ann Ferguson (eds). New York: Routledge, 129–44.
Curtin, Deane. 1991. "Toward an Ecological Ethic of Care." *Hypatia* 6 (1): 60–74.
—1992. "Recipe for Values." In *Cooking, Eating, Thinking,* D. Curtin and
     L. Heldke (eds). Bloomington: Indiana University Press, 123–44.
Damasio, Antonio R. 2010. *Self Comes to Mind: Constructing the Conscious Brain.*
     1st edn. New York: Pantheon Books.
Danford, Natalie. 2004. "DaCapo Embraces Metrosexuality." *Publisher's Weekly*
     (January 29): 107.
Davies, Sharyn Graham. 2006. *Challenging Gender Norms: Five Genders Among*
     Bugis in Indonesia. Independence, KY: Wadsworth Publishing.
Davion, Victoria. 1994. "Is Ecofeminism Feminist?" In *Ecological Feminism,*
     Karen J. Warren (ed.). New York: Routledge, 8–28.
Davis, Karen. 1988. "Farm Animals and the Feminine Connection." *Animals'*
     Agenda. 8 (Jan/Feb): 38–9.
—1995. "Thinking Like a Chicken: Farm Animals and the Feminine Connection."
     In *Animals and Women: Feminist Theoretical Explorations,* Carol J. Adams and
     Josephine Donovan (eds). Durham, NC: Duke University Press, 192–212.
—2001. *More than a Meal: The Turkey in History, Myth, Ritual, and Reality.* New
     York: Lantern Books.
Davis, Susan and Margo Demello. 2003. *Stories Rabbits Tell: A Natural and*
     Cultural History of a Misunderstood Creature. New York: Lantern Press.
Deckha, Maneesha. 2006. "The Salience of Species Difference for Feminist
     Theory." *Hastings Women's Law Journal* 17 (1): 1–38.
—2008. "Intersectionality and Post-human Visions of Equality. *Wisconsin Journal*
     of Law. 23.
—2012. "Toward a Postcolonial, Posthumanist Feminist Theory: Centralizing Race
     and Culture in Feminist Work on Nonhuman Animals." *Hypatia* 27 (3): 527–45.
Derrida, Jacques. 1991. "'Eating Well,' or the Calculation of the Subject." In *Who*
     Comes After the Subject?, Eduardo Cadava, Peter Conor, and Jean-Luc Nancy
     (eds). New York: Routledge.
—2002. "The Animal That Therefore I Am (More to Follow)." *Critical Inquiry* 28
     (Winter): 369–418.
Devall, Bill and George Sessions. 1984. *Deep Ecology: Living as if the Planet*
     Mattered. Salt Lake City, UT: Peregrine Smith Books.
Diamond, Cora. 1991a. "Eating Meat and Eating People." In *The Realistic Spirit.*
     Cambridge, MA: The MIT Press. 319–34.
—1991b. "Anything But Argument." In *The Realistic Spirit.* Cambridge, MA: The
     MIT Press, 335–51.
—2008. "The Difficulty of Reality." In *Philosophy and Animal Life.* New York:
     Columbia University Press, 43–89.

Dombrowski, Daniel. 1988. *Hartshorne and the Metaphysics of Animal Rights.* Albany, NY: State University of New York Press.

—2004. *Divine Beauty: The Aesthetics of Charles Hartshorne.* Nashville, TN: Vanderbilt University Press.

Donaldson, Sue and Will Kymlicka. 2011. *Zoopolis.* New York: Oxford University Press.

Donovan, Josephine. 1990. "Animal Rights and Feminist Theory." *Signs* 15 (2): 350–75. Reprinted in *Ecofeminism: Women, Animals, Nature* (1993), Greta Gaard (ed.). Philadelphia: Temple University Press.

—1995. "Comment on George's 'Should Feminists be Vegetarians?'" *Signs: Journal of Women in Culture and Society* 21 (1): 226–9.

—1996a. "Attention to Suffering: A Feminist Caring Ethic for the Treatment of Animals." *Journal of Social Philosophy* 27 (1): 81–102.

—1996b. "Ecofeminist Literary Criticism: Reading the Orange." *Hypatia* 11 (2): 161–84.

—2004. "'Miracles of Creation': Animals in J. M. Coetzee's Work." *Michigan Quarterly Review* 43 (1): 78–93.

—2006. "Feminism and the Treatment of Animals: From Care to Dialogue." *Signs* 31 (2): 305–29.

—2013. "The Voice of Animals: A Response to Recent French Care Theory in Animal Ethics." *Journal for Critical Animal Studies* 11 (1).

Donovan, Josephine and Carol J. Adams (eds). 1996. *Beyond Animal Rights: A Feminist Caring Ethic for the Treatment of Animals.* New York: Continuum.

—2007. *The Feminist Care Tradition in Animal Ethics.* New York: Columbia University Press.

—"Introduction." In *The Feminist Care Tradition in Animal Ethics*, Josephine Donovan and Carol J. Adams (eds). New York: Columbia University Press.

Driskill, Qwo-Li. 2004. "Stolen From Our Bodies: First Nations Two-Spirits/Queers and the Journey to a Sovereign Erotic." *Studies in American Indian Literatures* 16 (2): 50–64. doi:10.1353/ail.2004.0020.

Drucker, Peter. 1996. "'In the Tropics There Is No Sin': Sexuality and Gay–Lesbian Movements in the Third World." *New Left Review* 1 (218): 75–101.

Dundes, Alan. 1994. "Gallus As Phallus: A Psychoanalytic Cross-Cultural Consideration of the Cockfight As Fowl Play." *The Cockfight: A Casebook*, Alan Dundes (ed.). Madison: University of Wisconsin Press. 241–82.

Eaton, David. 2002. "Incorporating the Other: Val Plumwood's Integration of Ethical Frameworks." *Ethics and the Environment* 7, no. 2: 153–80.

Eddington, A. S. 1928. *The Nature of the Physical World.* New York: Macmillan.

Emmerman-Mazner, Karen. 2012. "Beyond the Basic/Nonbasic Interests Distinction: A Feminist Approach to Inter-species Moral Conflict and Moral Repair." PhD Diss., University of Washington.

Epprecht, Marc. 2008. *Heterosexual Africa?: The History of an Idea from the Age of Exploration to the Age of AIDS.* Athens, OH: Ohio University Press.

Erickson, Bruce. 2010. "'fucking close to water': Queering the Production of the Nation." In Catriona Mortimer-Sandilands and Bruce Erickson (eds). *Queer Ecologies: Sex, Nature, Politics, Desire.* Bloomington, IN: Indiana University Press, 309–30.

Evans, Arthur. 1978. *Witchcraft and the Gay Counterculture.* Boston: Fag Rag Books.

Evans, Rhonda, DeAnn Gauthier, and Craig Forsyth. 1998. "Dogfighting: Symbolic Expression and Validation of Masculinity." *Sex Roles* 39 (11/12): 825–38.

Farrell, Perry. 1990. "Of Course." On *Ritual de lo Habitual*. Warner Brothers.

Fearnley-Whittingstall, Hugh. 2007. *The River Cottage Meat Book*. Berkeley, CA: Ten Speed Press.

Fechner, Gustav. 1908. *Nanna: Oder Über das Seelenleben der Pflanzen* (1848). 4th edn. Hamburg and Leipzig: Leopold Voss.

—1946. *Religion of a Scientist*, ed. and trans. Walter Lowrie. New York: Pantheon.

Feminists for Animal Rights. 1994. "Guidelines for Starting a Program for Animals in Danger of Battering." Chapel Hll, NC: Feminists for Animal Rights.

Ferber, Abby. 2007. "The Construction of Black Masculinity: White Supremacy Now and Then." *Journal of Sport & Social Issues* 31 (1): 11–24.

Finn, John. 1998. "Masculinity and Earth Gods." *Certified Male: A Journal of Men's Issues* (Australia). Issue #9. http://www.certifiedmale.com.au/issue9/earthgod.htm (accessed September 26, 2012).

Fisher, Elizabeth. 1979. *Woman's Creation: Sexual Evolution and the Shaping of Society*. Garden City, NY: Anchor Press/Doubleday.

Fitzgerald, Amy. 2005. *Animal Abuse and Family Violence: Researching the Interrelationships of Abusive Power*. Lewiston, NY: Mellen Studies in Sociology, vol. 48.

Fleischer, Richard, dir. 1973. *Soylent Green*. Metro Goldwyn Meyer.

Flocker, Michael. 2003. *The Metrosexual Guide to Style: A Handbook for the Modern Man*. Cambridge, MA: DaCapo Press.

Flynn, Clifton P. 1999. "Exploring the Link between Corporal Punishment and Children's Cruelty to Animals." *Journal of Marriage and the Family* 61: 971–81.

—2000a. "Why Family Professionals Can No Longer Ignore Violence Toward Animals." *Family Relations*. 49 (1): 87–95.

—2000b. "Woman's Best Friend: Pet Abuse and the Role of Companion Animals in the Lives of Battered Women." *Violence Against Women*. 6 (2): 162–77.

Food and Agriculture Organization (FAO) of the United Nations. 2006. *Livestock's Long Shadow*. Rome: Livestock, Environment and Development Initiative.

Forsyth, Craig and Rhonda Evans. 1998. "Dogmen: The Rationalization of Deviance." *Society & Animals* 6 (3): 203–18.

Fraiman, Susan. "Pussy Panic versus Liking Animals: Tracking Gender in Animal Studies." *Critical Inquiry* 39 (1): 89–115.

Francione, Gary. 1996. "Ecofeminism and Animal Rights: A Review of *Beyond Animal Rights: A Feminist Caring Ethic for the Treatment of Animals*." *Women's Rights Law Reporter* 18: 95–106.

—2000. *Introduction to Animal Rights: Your Child or the Dog?* Philadelphia, Pennsylvania: Temple University Press.

—2004. "Animals—Property or Persons?" In *Animal Rights: Current Debates and New Directions*, Cass R. Sunstein and Martha Nussbaum (eds). Oxford: Oxford University Press.

—2009. "We're all Michael Vick." August 14. http://www.philly.com/dailynews/opinion/20070822_Were_all_Michael_Vick.html.

—2012. "Animal Rights: The Abolitionist Approach" (blog) http://www.abolitionistapproach.com/pets-the-inherent-problems-of-domestication/#.UmFyuuBkIqZ (accessed October 18, 2013).

Fraser, James A. and Joseph Heitman. 2004. "Evolution of Fungal Sex Chromosomes." *Molecular Microbiology* 51 (2): 299–306. doi:10.1046/j.1365-2958.2003.03874.x.

French, Marilyn. 1985. *Beyond Power: On Men, Women and Morals.* New York: Summit.

Frick, Robert. 2004. "The Manly Man's Guide to Makeup and Metrosexuality." *Kiplinger's Report* (June), 38.

Fry, Christopher. 1965. "Comedy." In *Comedy, Meaning and Form*, Robert W. Corrigan (ed.). San Francisco, CA: Chandler Publishing, 15–16.

Fudge, Erica. 2000. *Perceiving Animals: Humans and Beasts in Early Modern English Culture.* Champagne, IL: University of Illinois Press.

Gaard, Greta. 1993. "Ecofeminism and Native American Cultures: Pushing the Limits of Cultural Imperialism?" In *Ecofeminism: Women, Animals, Nature,* Greta Gaard (ed.). Philadelphia: Temple University Press.

—(ed.). 1993. *Ecofeminism: Women, Animals, Nature.* Philadelphia: Temple University Press.

—1994. "Milking Mother Nature: An Ecofeminist Critique of rBGH." *The Ecologist* 24:6. 1–2.

—1997. "Toward a Queer Ecofeminism." *Hypatia* 12 (1), 114–37.

—2001. "Tools for a Cross-Cultural Feminist Ethics: Exploring Ethical Contexts and Contents in the Makah Whale Hunt." *Hypatia* 16 (1): 1–26.

—2002. "Vegetarian Ecofeminism: A Review Essay." *Frontiers: A Journal of Women Studies* 23, no. 2: 117–46.

—2007. *Nature of Home: Taking Root in a Place.* Tucson: University of Arizona Press.

—2011. "Ecofeminism Revisited: Rejecting Essentialism and Re-Placing Species in a Material Feminist Environmentalism." *Feminist Formations* 23 (2): 26–53.

Gaard, Greta and Lori Gruen. 1993. "Ecofeminism: Toward Global Justice and Planetary Health. *Society and Nature* 2 (1): 1–35.

—1995. "Comment on George's 'Should Feminists be Vegetarians?'." *Signs: Journal of Women in Culture and Society* 21 (1): 230–41.

Galeano, Eduardo. 1992. *We Say No.* New York: W. W. Norton & Company.

Garland-Thomson, Rosemarie. 2009. *Staring: How We Look.* New York and London: Oxford University Press.

George, Kathryn P. 1994. "Should Feminists be Vegetarians?" *Signs: Journal of Women in Culture and Society* 19 (2): 405–34.

—1995. "Reply to Adams, Donovan, and Gaard and Gruen." *Signs: Journal of Women in Culture and Society* 21 (1): 242–60.

—2000. *Animal, Vegetable, or Woman? A Feminist Critique of Ethical Vegetarianism.* Albany, NY: SUNY Press.

Gibson, Hanna. 2005. "Dog Fighting Detailed Discussion." Animal Legal and Historical Center, Michigan State University College of Law. http://www.animallaw.info/articles/ddusdogfighting.htm.

Gilbert, Sandra. 2006. *Death's Door: Modern Dying and the Ways we Grieve.* New York: W. W. Norton & Company.

Gilligan, Carol. 1982. *In a Different Voice: Psychological Theory and Women's Development.* Cambridge: Harvard University Press.

Gilmore, Ruth Wilson. 2007. *Golden Gulag: Prisons, Surplus, Crisis, and Opposition in Globalizing California.* Berkeley: University of California Press.

Gorant, Jim. 2010. *The Lost Dogs: Michael Vick's Dogs and Their Tale of Rescue and Redemption*. New York: Gotham Books.

Gordon-Reed, Annette. 2009. *The Hemingses of Monticello: An American Family*. New York: W. W. Norton & Company.

Gramsci, Antonio. 1971. *Selections from the Prison Notebooks*, edited by Quintin Hoare and Geoffrey Nowell Smith. New York: International Publishers Co.

Grandin, Temple and Catherine Johnson. 2009. *Animals Make Us Human: Creating the Best Life for Animals*. Boston: Houghton Mifflin Harcourt.

Gray, Elizabeth Dodson. 1979. *Green Paradise Lost*. Wellesley, MA: Roundtable Press.

Griffin, David Ray. 1998. *Unsnarling the World-Knot: Consciousness, Freedom, and the Mind-Body Problem*. Berkeley: University of California Press.

Gruen, Lori. 1991. "Animals." In *A Companion to Ethics*, Peter Singer (ed.). Oxford: Blackwell Publishers.

—1993. "Dismantling Oppression: An Analysis of the Connection between Women and Animals." In *Ecofeminism: Women, Animals, Nature,* Greta Gaard (ed.). Philadelphia: Temple University Press.

—1994. "Toward an Ecofeminist Moral Epistemology." In *Ecological Feminism,* Karen Warren (ed.). New York: Routledge.

—1996. "On The Oppression of Women and Animals." *Environmental Ethics* 18 (4): 441–4.

—2001. "Beyond Exclusion: The Importance of Context in Ecofeminist Theory." *Land, Value, Community,* J. Hall and W. Ouderkirk (eds). Albany, NY: SUNY Press.

—2002. "Conflicting Values in a Conflicted World." *Women & Environments International,* #52/53: 16–18.

—2004. "Empathy and Vegetarian Commitments." In *Food for Thought*, Steve F. Sapontzis (ed.). Amherst, NY: Prometheus Books.

—2009. "Attending to Nature: Empathetic Engagement with the More Than Human World." *Ethics & the Environment* 14 (2): 23–38.

—2012. "Navigating Difference (Again): Animal Ethics and Entangled Empathy." In *Strangers to Nature: Animal Lives and Human Ethics*, G. Zucker (ed.). New York: Lexington Books.

—2013. "Entangled Empathy: An Alternate Approach to Animal Ethics." In *The Politics of Species: Reshaping our relationships with other animals*, Raymond Corbey and Annette Lanjouw (eds). New York: Cambridge University Press.

—(ed.). 2014. *The Ethics of Captivity*. New York: Oxford University Press.

Gruen, Lori and Kari Weil. 2012. "Animal Others—Editors' Introduction." Special Issue, *Hypatia* 27 (3): 477–87.

Guthman, J. 2008. "'If They Only Knew': Color Blindness and Universalism in California Alternative Food Institutions." *The Professional Geographer* 60 (3): 387–97.

Halberstam, Judith. 1998. *Female Masculinity*. Durham, NC: Duke University Press.

Hale, Grace Elizabeth. 1998. *Making Whiteness: The Culture of Segregation in the South: 1890–1940*. New York: Vintage Books.

Haraway, Donna. 1988. *Primate Visions*. New York: Routledge.

—1997. *Modest_Witness@Second_Millennium: FemaleMan©– Meets– OncoMouse™*. New York: Routledge.

—2003. *The Companion Species Manifesto: Dogs, People, and Significant Otherness.* Prickly Paradigm Press.

—2008. *When Species Meet.* Minneapolis: University of Minnesota Press.

Harper, A. Breeze. 2010a. "Race as a 'Feeble Matter' in Veganism: Interrogating whiteness, geopolitical privilege, and consumption philosophy of 'cruelty-free' products." *Journal for Critical Animal Studies* 8 (3): 5–27.

—2010b. *Sistah Vegan: Black Female Vegans Speak on Food, Identity, Health, and Society.* New York: Lantern Books.

Harris, Michael. 2003. *Colored Pictures: Race and Representation.* Chapel Hill and London: The University of North Carolina Press.

Harrison, Harry. 1966. *Make Room! Make Room!* New York: Doubleday.

Hartshorne, Charles. 1937. *Beyond Humanism: Essays in a New Philosophy of Nature.* Chicago: Willett, Clark.

—1973. *Born to Sing: An Interpretation and World Survey of Bird Song.* Bloomington: Indiana University Press.

—1979. "The Rights of the Subhuman World." *Environmental Ethics* 1 (Spring): 49–60.

—1980. "In Defense of Wordsworth's View of Nature." *Philosophy and Literature* 4, no. 1 (Spring): 80–91.

Harvey, David. 1997. *The Condition of Postmodernity.* Cambridge, MA: Blackwell.

Hassan, Ihab. 1985. "The Culture of Postmodernism." *Theory, Culture, and Society*, 2 (3): 119–32.

Hawkins, R. 2001. "Cultural whaling, commodification, and culture change." *Environmental Ethics* 23 (3): 287–306.

Heisenberg, Werner. 1958. *Physics and Philosophy: The Revolution in Modern Science.* New York: Harper & Row.

Held, Virginia. 1995. "Feminist Moral Inquiry and the Feminist Future." In *Justice and Care: Essential Readings in Feminist Ethics*, Virginia Held (ed.). Boulder, CO: Westview Press.

Heller, Chaia. 1999. *Ecology of Everyday Life: Rethinking the Desire for Nature.* Montreal: Black Rose Books.

Hird, Myra J. 2004. "Naturally Queer." *Feminist Theory* 5: 85–9.

Hoberman, John. 1997. *Darwin's Athletes: How Sport Has Damaged Black America and Preserved the Myth of Race.* Boston: Houghton Mifflin.

Hughes, Robert. 1980. *The Shock of the New: Modern Art, Its Rise, Its Dazzling Achievement, Its Fall.* New York: Alfred A. Knopf.

Hurn, Samantha. 2012. *Humans and Other Animals: Cross-Cultural Perspectives on Human-Animal Interactions.* London: Pluto Press.

Hursthouse, Rosalind. 2006. "Applying Virtue Ethics to Our Treatment of Other Animals." In *The Practice of Virtue*, Jennifer Welchman (ed.). Indianapolis, IN: Hackett Publishing.

Jackson, Peter A. 2001. "Pre-Gay, Post-Queer: Thai Perspectives on Proliferating Gender/ Sex Diversity in Asia." *Journal of Homosexuality*, 40 (3/4): 1–25

Jefferson, Thomas. 1794. *Notes on the State of Virginia.* Philadelphia: Printed for Mathew Carey, no. 118, Market-Street, November 12.

Jewett, Sarah Orne. 1881. "A Winter Drive." In *Country By-Ways.* Boston: Houghton Mifflin.

Jones, Keithly G. 2004. "Trends in the U.S. Sheep Industry." *Electronic Report from the Economic Research Service.* United States Department of Agriculture. http://usda.mannlib.cornell.edu/usda/ers/sheeptrends/aib787.pdf (accessed December 2, 2011).

Jordan, Winthrop. 1968. *White Over Black: American Attitudes Toward the Negro, 1550–1812.* Chapel Hill: University of North Carolina Press.

Kafer, Alison. 2013. *Feminist, Queer, Crip.* Bloomington, IN: Indiana University Press.

Kao, G. Y. 2010. "The Universal versus the Particular in Ecofeminist Ethics." *Journal of Religious Ethics* 38 (4): 616–37.

Kappeler, Susanne. 1986. *The Pornography of Representation.* Minneapolis: The University of Minnesota Press.

—1995. "Speciesism, Racism, Nationalism… or the Power of Scientific Subjectivity." In *Animals and Women: Feminist Theoretical Explorations,* Carol J. Adams and Josephine Donovan (eds). Durham, NC: Duke University Press.

Karpova, Lisa. (trans.) 2011. "Third Sex Prehistoric Skeleton Found." *Pravda,* July 4. http://english.pravda.ru/science/tech/07-04-2011/117498-Third_ Sex_prehistoric_skeleton_found-0/.

Katyal, Sonia. 2002. "Exporting Identity." *Yale Journal of Law and Feminism* 14 (1). http://papers.ssrn.com/sol3/Delivery.cfm/SSRN_ID330061_code021111140.pdf

Katz, Jonathan. 1976. *Gay American History.* New York: Avon.

—2007. *The Invention of Heterosexuality.* Chicago and London: University of Chicago Press.

Kean, Hilda. 1995. "The 'Smooth Cool Men of Science': The Feminist and Socialist Response to Vivisection." *History Workshop Journal.* 40 (1): 16–38.

Keller, Evelyn Fox. 1983. *A Feeling for the Organism: The Life and Work of Barbara McClintock.* San Francisco: W. H. Freeman.

Kelly, Christine. 2013. "Building Bridges with Accessible Care: Disability Studies, Feminist Care, Scholarship, and Beyond." *Hypatia* 28 (4): 784–800.

Kemmerer, Lisa. 2011. *Sister Species: Women, Animals, and Social Justice.* Urbana, Chicago, and Springfield: University of Illinois Press.

Khan, S. 2001. "Culture, Sexualities, and Identities: Men Who Have Sex with Men in India." *Journal of Homosexuality* 40 (3–4): 99–115. doi:10.1300/J082v40n03_06.

Kheel, Marti. c. 1984. "A Feminist View of Mobilization." *Feminists for Animal Rights Newsletter* [First newsletter, no volume or issue number]: 2.

—1985. "The Liberation of Nature: A Circular Affair." *Environmental Ethics* 7 (2): 135–49.

—1990. "Ecofeminism and Deep Ecology: Reflections on Identity and Difference." In *Reweaving the World: The Emergence of Ecofeminism,* Irene Diamond and Gloria Feman Orenstein (eds). San Francisco: Sierra Club Books.

—1993. "From Heroic to Holistic Ethics: The Ecofeminist Challenge." In *Ecofeminism: Women, Animals, Nature,* Greta Gaard (ed.). Philadelphia: Temple University Press.

—2004. "Vegetarianism and Ecofeminism: Toppling Patriarchy with a Fork." In *Food for Thought: The Debate Over Eating Meat,* S. Sapontzis (ed.). Amherst, NY: Prometheus Books.

—2008. *Nature Ethics: An Ecofeminist Perspective.* Lanham, MD: Rowman & Littlefield Publishing Group, Inc.

Kim, Claire Jean. 2007. "Multiculturalism goes Imperial—Immigrants, Animals, and the Suppression of Moral Dialogue." *Du Bois Review* 4 (1): 233–49.

—2010. "Slaying the Beast: Reflections on Race, Culture, and Species." *Kalfou* 1 (1): 57–74.

—2011. "President Obama and the Polymorphous 'Other' in U.S. Racial Discourse." *Asian American Law Journal,* 18: 165–75.

—March 8, 2012. "The Great Yellow Hope." http://www.wbez.org/blog/alison-cuddy/2012-03-08/jeremy-lin-great-yellow-hope-97098.

King, Barbara. 2013. *How Animals Grieve.* Chicago: University of Chicago Press.

Kittay, Eva Feder. 2002. "Love's Labor Revisited." *Hypatia* 17 (3): 2327–50.

Kittay, Eva Feder and Diana T. Meyers (eds). 1987. *Women and Moral Theory.* Totowa, NJ: Rowman and Littlefield.

Kristof, Nicholas and Sheryl WuDunn. 2010. *Half the Sky: Turning Oppression into Opportunity Worldwide.* New York: Vintage.

Krupar, Shiloh R. 2012. "Transnatural Ethics: Revisiting the Nuclear Cleanup of Rocky Flats, CO, Through the Queer Ecology of Nuclia Waste." *Cultural Geographies* 19 (3): 303–27.

Laclau, Ernesto and Chantal Mouffe. 1985. *Hegemony and Socialist Strategy: Toward a Radical Democratic Politics.* London: Verso.

Lama, Dalai. 2011. *Beyond Religion: Ethics for a Whole World.* Boston: Houghton Mifflin Harcourt.

—n.d. "Compassion and the Individual." http://www.dalailama.com/messages/compassion.

Langer, Susanne. 1965. "The Comic Rhythm." In *Comedy, Meaning and Form,* Robert W. Corrigan (ed.). San Francisco, CA: Chandler Publishing.

Lansbury, Coral. 1985. *The Old Brown Dog: Women, Workers and Vivisection in Edwardian England.* Madison: The University of Wisconsin Press.

Lappé, Frances Moore. 1972. *Diet for a Small Planet.* New York: Ballantine Books.

Larabee, Mary Jeanne (ed.) 1993. *An Ethic of Care: Feminist and Interdisciplinary Perspectives.* New York: Routledge.

Laucella, Pamela. 2010. "Michael Vick: An Analysis of Press Coverage on Federal Dogfighting Charges. *Journal of Sports Media* 5 (2): 35–76.

Lederman, Leon M. and Christopher T. Hill. 2011. *Quantum Physics for Poets.* Amherst, NY: Prometheus.

Leonard, David. 2010. "Jumping the Gun: Sporting Cultures and the Criminalization of Black Masculinity." *Journal of Sport & Social Issues* 34 (2): 252–62.

Leonard, David and C. Richard King (eds). 2011. *Commodified and Criminalized: New Racism and African Americans in Contemporary Sports.* Lanham, MD: Rowman & Littlefield.

Lockwood, Randall and Frank R. Ascione. 1998. *Cruelty to Animals and Interpersonal Violence: Readings in Research and Application.* West Lafayette, IN: Purdue University Press.

Lorde, Audre. 2012. *Sister Outsider: Essays and Speeches.* New York: Random House.

Lucas, S. 2005. "A Defense of the Feminist-Vegetarian Connection." *Hypatia* 20 (1): 150–77.

Luke, Brian. 1995. "Taming Ourselves or Going Feral? Toward a Nonpatriarchal Metaethic for Animal Liberation." In *Animals and Women: Feminist Theoretical Explorations,* Carol J. Adams and Josephine Donovan (eds). Durham: Duke University Press.

—1996. "The Erotics of Predation: An Ecofemnist Look at *Sports Illustrated.*" *Feminists for Animal Rights Newsletter* X (1–2): 6–7.

—1997. "A Critical Analysis of Hunters' Ethics." *Environmental Ethics* 19 (Spring): 25–44.

—1998. "Violent Love: Hunting, Heterosexuality, and the Erotics of Men's Predation." *Feminist Studies* 24 (Fall): 627–55.

—2007. "Justice, Caring, and Animal Liberation." In *The Feminist Care Tradition in Animal Ethics,* Josephine Donovan and Carol J. Adams (eds). New York: Columbia University Press.

Marder, Michael. 2012. "If Peas Can Talk, Should We Eat Them?" *New York Times,* April 29. "Sunday Review," 9.

Margulis, Lynn. 2001. "The Conscious Cell." *Annals of the New York Academy of Sciences* 929: 55-70.

Martindale, Steve. 2010. "What's Wrong with Wool? http://prime.peta.org/2010/05/whats-wrong-with-wool (accessed December 2, 2011).

Mason, Jim. 1981. "Fear and Loathing on the Hog Farmer Trail: Jim Mason takes a poke at the 1981 American Pork Congress." *Vegetarian Times* (June) 47: 66–8.

—1997. *An Unnatural Order: Why We Are Destroying the Planet and Each Other.* New York: Continuum.

Mason, Jim and Peter Singer. 1980. *Animal Factories: The Mass Production of Animals For Food and How It Affects The Lives of Consumers, Farmers, and the Animals Themselves.* New York: Crown.

Massey, D. 1991. "Flexible Sexism." *Environment and Planning* 9: 31–57.

Mathews, Freya. 2003. *For Love of Matter: A Contemporary Panpsychism.* Albany, NY: State University of New York Press.

McAllister, Pam. 1982. "Introduction to article by Connie Salamone." *Reweaving the Web of Life: Feminism and Nonviolence.* Philadelphia: New Society Publishers.

McClintock, Anne. 1995. *Imperial Leather: Race, Gender and Sexuality in the Colonial Context.* New York: Routledge.

McComb, Karen, Lucy Baker, and Cynthia Moss. 2006. "African elephants show high levels of interest in the skulls and ivory of their own species." *Biology Letters* 2 (1): 26–8.

McKittrick, Katherine. 2006. *Demonic Grounds: Black Women and the Cartographies of Struggle.* Minneapolis: University of Minnesota Press.

McWhorter, Ladelle. 2010. "Enemy of the Species." In *Queer Ecologies: Sex, Nature, Politics, Desire*, Catriona Mortimer-Sandilands and Bruce Erickson (eds). Bloomington, IN: Indiana University Press. 73–101.

Mellon, Joseph. 1989. *Nature Ethics without Theory.* PhD Diss., University of Oregon.

Mellor, Mary. 1992. *Breaking the Boundaries: Towards a Feminist Green Socialism.* London: Virago Press.

—1997. *Feminism & Ecology.* Washington Square, NY: New York University Press.

Melville, Herman. 1964. *Moby Dick.* New York: Bantam.

Merchant, Carolyn. 1980. *The Death of Nature: Women, Ecology and the Scientific Revolution*. New York: Harper & Row.

Metzinger, Thomas. 2009. *The Ego Tunnel: The Science of the Mind and the Myth of the Self*. New York: Basic Books.

Meyerding, Jane. 1982. "Feminist Criticism and Cultural Imperialism (Where does one end and the other begin)." In *Animals Agenda* (November/December) 14–15, 22–3.

"The Michael Vick Case: Is 'Supporting Our Own' OK?" August 22, 2007. *Tell Me More*, National Public Radio. http://www.npr.org/templates/story/story.php?storyId=13859969.

Morgan, Ruth and Saskia Wierenga. 2006. *Tommy Boys, Lesbian Men, and Ancestral Wives: Female Same-Sex Practices in Africa*. Johannesburg: Jacana Media.

Morrison, Toni. 1992. *Playing in the Dark: Whiteness and the Literary Imagination*. New York: Vintage Books.

Mortimer-Sandilands, Catriona and Bruce Erickson. 2010. "A Genealogy of Queer Ecologies." In *Queer Ecologies: Sex, Nature, Politics, Desire*, Catriona Mortimer-Sandilands and Bruce Erickson (eds). Bloomington, IN: Indiana University Press. 1–47.

—(eds). 2010. *Queer Ecologies: Sex, Nature, Politics, Desire*. Bloomington, IN: Indiana University Press.

Muhammad, Latifah. 2012. "Police Raid Rapper Young Calicoe's Home Over Suspected Dog-Fighting." July 11. http://hiphopwired.com/2012/07/11/.police-raid-rapper-young-calicoes-home-over-suspected-dog-fighting-video/

Murdoch, Iris. 2001. *The Sovereignty of Good*. 2nd edn. New York: Routledge.

Murphy, Patrick D. 1995. *Literature, Nature, and Other: Ecofeminist Critiques*. Albany, NY: State University of New York Press.

—2004. "Nature Nurturing Fathers in a World Beyond Our Control." In *Eco-Man: New Perspectives on Masculinity and Nature*, Mark Allister (ed.). Charlottesville: University of Virginia Press. 196–210.

Nagel, Thomas. 2012. *Mind and Cosmos: Why the Materialist Neo-Darwinian Conception of Nature is Almost Certainly False*. New York: Oxford University Press.

—2013. "The Core of 'Mind and Cosmos.'" *New York Times*, August 18.

Nanda, Serena. 2000. *Gender Diversity: Crosscultural Variations*. Long Grove, IL: Waveland Press.

Nash, Roderick Frazier. 1989. *The Rights of Nature: A History of Environmental Ethics*. Madison: University of Wisconsin Press.

Noddings, Nel. 1984. *Caring: A Feminine Approach to Ethics and Moral Education*. Berkeley: University of California Press.

Noland, William F. and George Clayton Johnson. 1967. *Logan's Run*. NYC: Dial Press.

Noske, Barbara. 1989. *Humans and Other Animals: Beyond the Boundaries of Anthropology*. London: Pluto Press.

Nott, Josiah and George R. Gliddon. 1855. *Types of Mankind: Or, Ethnological Researches: Based Upon the Ancient Monuments, Paintings, Sculptures, and Crania of Races, and Upon Their Natural, Geographical, Philological and Biblical History, Illustrated by Selections from the Inedited Papers of Samuel George Morton and by Additional Contributions from L. Agassiz, W. Usher, and H. S. Patterson*. London, Trübner & Co.

Nussbaum, Martha C. 1992. *Love's Knowledge: Essays in Philosophy and Literature*. New York: Oxford University Press.

—2000. *Women and Human Development: The Capabilities Approach, The John Robert Seeley lectures*. Cambridge; New York: Cambridge University Press.

—2001. *Upheavals of Thought: The Intelligence of Emotions*. Cambridge and New York: Cambridge University Press.

—2004. "Beyond 'Compassion and Humanity?': Justice for Nonhuman Animals." In *Animal Rights: Current Debates and New Directions*, Cass Sunstein and Martha C. Nussbaum (eds). Chicago: University of Chicago Press. 299–320.

—2006. *Frontiers of Justice: Disability, Nationality, Species Membership*. Cambridge, MA: Harvard University Press, 2006.

—2009. "Justice." In *Examined Life: Excursions with Contemporary Thinkers*, Astra Taylor (ed.). New York: New Press.

Oates, Thomas Patrick. 2007. "The Erotic Gaze in the NFL Draft." *Communication and Critical/Cultural Studies* 4 (1): 74–90.

Oliver, Michael. 1990. *The Politics of Disablement: A Sociological Approach*. New York: St Martin's Press.

Ornelas, lauren. 2012. "Maturing of Activism." Paper presented at "Finding a Niche for All Animals: A Conference Honoring the Ecofeminist Work of Marti Kheel." Wesleyan University.

Orr, H. Allen. 2013. "'Awaiting a New Darwin' (review of Thomas Nagel, *Mind and Cosmos*)." *New York Review of Books*, February 7, 26–8.

Oscamp, S. 2000. "A sustainable future for humanity? How can psychology help?" *American Psychologist*. 55(5): 496–508.

Oyama, Susan. 2000. *The Ontogeny of Information: Developmental Systems and Evolution*. Durham and London: Duke University Press.

Painter, Nell Irvin. 2010. *The History of White People*. New York: W. W. Norton & Company.

Palmer, C. 2010. *Animal Ethics in Context*. New York: Columbia University Press.

Patterson, Charles. 2002. *Eternal Treblinka: Our Treatment of Animals and the Holocaust*. New York: Lantern Books.

Penrose, Roger. 1994. *Shadows of the Mind: A Search for the Missing Science of Consciousness*. Oxford: Oxford University Press.

PETA Files. 2007. Vick at the Office, Part 2. http://blog.peta.org/archives/2007/10/vick_at_the_off.php.

Peters, Sharon. 2008. "A Fight to Save Urban Youth from Dogfighting." *USA TODAY.com*. September 29. http://usatoday30.usatoday.com/news/nation/2008-09-29-dogfighting_N.htm.

Pharr, Suzanne. 1988. *Homophobia: A Weapon of Sexism*. Little Rock, AR: Women's Project Press.

Pluhar, Evelyn. 1995. *Beyond Prejudice: The Moral Significance of Human and Nonhuman Animals*. Durham and London: Duke University Press.

—2006. "Legal and Moral Rights of Sentient Beings and the Full Personhood View." *Organization and Environment* 19 (2): 275–8.

Plumwood, Val. 1993. *Feminism and the Mastery of Nature*. London and New York: Routledge.

—1995. "Human Vulnerability and the Experience of Being Prey." *Quadrant* March: 29–34.

—2000. "Integrating Ethical Frameworks for Humans, Animals, and Nature: A Critical Feminist Eco-Socialist Perspective." *Ethics and the Environment* (Spring).

—2001. *Environmental Culture: The Ecological Crisis of Reason*. London: Routledge.

—2003. "Animals and Ecology: Towards a Better Integration." Working/ Technical Paper, *Australian National University Digital Collection*: https:// digitalcollections.anu.edu.au/handle/1885/41767

—2004. "Gender, Eco-Feminism and the Environment." In *Controversies in Environmental Sociology*, R. White (ed.). Cambridge: Cambridge University Press. 43–60.

Pollan, Michael. 2009. *Omnivore's Dilemma: A Natural History of Four Meals*. New York: Penguin Group.

Potts, Annie and Jovian Parry. 2010. "Vegan Sexuality: Challenging Heteronormative Masculinity through Meat-free Sex." *Feminism & Psychology* 20 (1): 53–72.

Pulé, Paul M. 2007. "Ecology and Environmental Studies." In *Routledge International Encyclopedia of Men and Masculinities,* Michael Flood, Judith Kegan Gardiner, Bob Pease, and Keith Pringle (eds). New York: Routledge, 158–62.

—2009. "Caring for Society and Environment: Towards Ecological Masculinism." Paper Presented at the Villanova University Sustainability Conference, April 2009. http://www.paulpule.com.au/Ecological_Masculinism.pdf (accessed October 1, 2012).

de Quincey, Christian. 2002. *Radical Nature: Rediscovering the Soul of Matter*. Montpelier, VT: Invisible Cities Press.

"Race Played Factor in Vick Coverage, Critics Say." 2007. Neal Conan's *Talk of the Nation*, National Public Radio, August 28. http://www.npr.org/templates/story/ story.php?storyId=14000094.

"Racist Obama Email: Marilyn Davenport Insists It Was Satire." 2011. *Huffington Post* April 20. http://www.huffingtonpost.com/2011/04/20/racist-obama-email-marilyn-davenport_n_851772.html.

Ramet, Sabrina Petra (ed.). 1996. *Gender Reversals and Gender Cultures: Anthropologial and Historical Perspectives*. London: Routledge.

Regan, Tom. 1983. *The Case for Animal Rights*. Berkeley: University of California Press.

Regan, Tom and Peter Singer. 1989. *Animal Rights and Human Obligations*. 2nd edn. Englewood Cliffs, NJ: Prentice Hall.

Ricciardelli, Rosemary, Kimberley A. Clow, and Philip White. 2010. "Investigating Hegemonic Masculinity: Portrayals of Masculinity in Men's Lifestyle Magazines." *Sex Roles* 63: 64–78.

"Richard Cebull, Montana Federal Judge, Admits Forwarding Racist Obama Email." 2012. *Huffington Post* March 1. http://www.huffingtonpost. com/2012/03/01/richard-cebull-judge-obama-racist- email_n_1312736.html.

Ristau, Carolyn A. 2013. "Cognitive Ethology." *WIREs Cognitive Science* 4 (September–October): 493–509.

Ritvo, Harriet. 1998. *The Platypus and the Mermaid: And Other Figments of the Classifying Imagination*. Cambridge: Harvard University Press.

Roediger, David R. 2008. *How Race Survived U.S. History: From Settlement and Slavery to the Obama Phenomenon*. London and New York: Verso.

Roscoe, Will (ed.). 1988. *Living the Spirit: A Gay American Indian Anthology*. New York: St. Martin's Press.

—2000. *Changing Ones: Third and Fourth Genders in Native North America*. New York: Palgrave Macmillan.

Rose, Deborah Bird. 2013. "In the Shadow of All This Death." In *Animal Death*, Jay Johnston and Fiona Probyn-Rapsey (eds). Sydney: Sydney University Press.

Rubin, James H. 1994. *Manet's Silence and the Poetics of Bouquets*. London: Reaktion Books.

Ruddick, Sara. 1989. *Maternal Thinking: Toward a Politics of Peace*. Boston: Beacon.

Ruether, Rosemary Radford. 1974. *New Woman/New Earth: Sexist Ideologies and Human Liberation*. New York: Seabury.

—1983. *Sexism and God-Talk: Toward a Feminist Theology*. Boston: Beacon Press.

—1992. *Gaia & God: An Ecofeminist Theology of Earth Healing*. San Francisco: HarperCollins.

Salamone, Connie. 1973. "Feminist as Rapist in the Modern Male Hunter Culture." *Majority Report* (October).

—1982. "The Prevalence of the Natural Law within Women: Women and Animal Rights." In *Reweaving the Web of Life: Feminism and Nonviolence,* Pam McAllister (ed.). Philadelphia: New Society Publishers.

Salleh, Ariel. 1984. "Deeper Than Deep Ecology: The Eco-Feminist Connection." *Environmental Ethics* 6: 339–45.

—1997. *Ecofeminism as Politics: Nature, Marx, and the Postmodern*. London: Zed Books.

Sanbonmatsu, John. 2005. "Listen, Ecological Marxist! (Yes, I said *Animals*!)." *Capitalism Nature Socialism* 16 (2): 107–14.

Sanders, Douglas. 2005. "Flying the Rainbow Flag in Asia." Conference paper. http://digitalcollections.anu.edu.au/handle/1885/8691

Sandilands, Catriona. 1994. "Lavender's Green? Some Thoughts on Queer(y)ing Environmental Politics." *Undercurrents: Critical Environmental Studies* (May): 20–4.

Śāntideva, Kate Crosby and Andrew Skilton. 2008. *The Bodhicaryāvatāra, Oxford World's Classics*. Oxford; New York: Oxford University Press.

Sapontzis, Steve (ed.). 2004. *Food for Thought: The Debate Over Eating Meat*. Amherst, NY: Prometheus Books.

Scheler, Max. 1970. *The Nature of Sympathy* (1923), trans. Peter Heath. Hamden, CT: Archon.

Schwalbe, Michael. 2012. "The Hazards of Manhood." *Yes!* 63 (Fall): 42–4.

Scully, J. L., L. Baldwin-Ragaven, and P. Fitzpatrick (eds). 2010. *Feminist Bioethics: At the Center, on the Margins*. Baltimore: Johns Hopkins University Press.

Seaberg, Kurt. 2010. "Artist's Statement." http://www.kurtseaberg.com/statement.html (accessed October 22, 2012).

Shapiro, Kenneth. 1989. "Understanding Dogs through Kinesthetic Empathy, Social Construction, and History." *Anthrozoös* 3, no. 3: 184–95.

Shove, E. and M. Pantzar. 2005. "Consumers, Producers and Practices— Understanding the invention and reinvention of Nordic walking." *Journal of Consumer Culture* 5 (1): 43–64.

Shukin, Nicole. 2009. *Animal Capital: Rendering Life in Biopolitical Times*. Minneapolis: University of Minnesota Press.

Simonet, Patricia. 2007. "Laughing Dogs." *The Bark*, 44: 2–3.

Singer, June. 1977. *Androgyny: Toward a New Theory of Sexuality*. New York: Anchor Books/Doubleday.

Singer, Peter. 1990. *Animal Liberation*. Revised edition. New York: Avon Books.

—1994. "Feminism and Vegetarianism: A Response." *Philosophy in the Contemporary World* 1 (2): 36–9.

—2009. *Animal Liberation: The Definitive Classic of the Animal Movement*. Reissue edn. New York: Harper Perennial Modern Classics.

Skolimowki, Henryk. 1994. *The Participatory Mind: A New Theory of Knowledge and the Universe*. London: Penguin Arkana.

Skrbina, David. 2005. *Panpsychism in the West*. Cambridge, MA: The MIT Press.

Slicer, Deborah. 1991. "Your Daughter or Your Dog: A Feminist Assessment of the Animal Research Issue." *Hypatia* 6 (1): 108–24.

Smith, Andrea. 2005. *Conquest: Sexual Violence and American Indian Genocide*. Cambridge, MA: South End Press.

Smith, Cheryl. 2010. "The Little Dog Laughed—The Function and Form of Dog Play." *IAABC Dog Blog*, December 6.

Smuts, Barbara. 2008. "Reflections." In *The Lives of Animals* by J. M. Coetzee. Princeton, NJ: Princeton University Press, 107–20.

Smythe, R. H. 1965. *The Mind of the Horse*. Brattleboro, VT: The Stephen Greene Press.

Solomon, Robert. 1976. *Passions*. Garden City, NY: Anchor Press/Doubleday.

Sorabji, Richard. 1993. *Animal Minds and Human Morals: The Origins of the Western Debate*. Ithaca, NY: Cornell University Press.

Spät, Patrick. 2009. "Panpsychism, the Big-Bang-Argument, and the Dignity of Life." In *Mind that Abides: Panpsychism in the New Millennium*, David Skrbina (ed.). Amsterdam: John Benjamins, 159–76.

Spencer, C. 2000. *Vegetarianism: A History*. London, Grub Street.

Spiegel, Marjorie. 1996. *The Dreaded Comparison: Human and Animal Slavery*. Mirror Books.

de Spinoza, Benedictus and Michael L. Morgan. 2006. *The Essential Spinoza: Ethics and Related Writings*. Indianapolis: Hackett Publishing.

Spretnak, Charlene (ed.). 1982. *The Politics of Women's Spirituality*. New York: Anchor Books.

Sprinkle, Annie. 1991. "Beyond Bisexual." In *Bi Any Other Name: Bisexual People Speak Out,* edited by Loraine Hutching and Lani Kaahumanu. New York: Alyson Publications, 103–7.

Stanescu, James. 2012. "Species Trouble: Judith Butler, Mourning, and the Precarious Lives of Animals." *Hypatia* 27 (3): 567–82.

Steeves, H. Peter (ed.). 1999. *Animal Others: On Ethics, Ontology, and Animal Life*. Albany: SUNY Press.

Stein, Edith. 1966 (1916). *On the Problem of Empathy*. The Hague: Martinus Nijhoff.

Steiner, Gary. 2010. *Anthropocentrism and Its Discontents: The Moral Status of Animals in the History of Western Philosophy*. Pittsburgh: University of Pittsburgh Press.

Steingraber, Sandra. 1997. *Living Downstream: An Ecologist Looks at Cancer and the Environment*. Reading, MA: Addison-Wesley.

Strouse, Kathy, with Dog Angel. 2009. *Bad Newz: The Untold Story of the Michael Vick Dogfighting Case*. Dogfighting Investigation Publications.

Swift, Jonathan. 1729. "A Modest Proposal." In *Encyclopedia of the Self*, edited by Mark Zimmerman. http://emotionalliteracyeducation.com/classic_books_online/mdprp10.htm (accessed February 9, 2013).

Tamale, Sylvia. 2007. "Out of the Closet: Unveiling Sexuality Discourses in Uganda." In *Africa after Gender?*, Catherine M. Cole, Takyiwaa Manuh, and Stephan F. Miescher (eds). Bloomington, IN: Indiana University Press, 17–29.

Tarter, Jim. 2002. "Some Live More Downstream than Others: Cancer, Gender, and Environmental Justice." In *The Environmental Justice Reader: Politics, Poetics, and Pedagogy*, Joni Adamson, Mei Mei Evans, and Rachel Stein (eds). Tucson, AZ: University of Arizona Press, 213–28.

Taylor, Paul. 1986. *Respect for Nature: A Theory of Environmental Ethics*. Princeton, NJ: Princeton University Press.

Taylor, Sunaura. 2004. "Right Not to Work: Power and Disability." *Monthly Review* 55, (10).

—2011. "Beasts of Burden: Disability Studies and Animal Rights." *Qui Parle* 19 (2).

Thomas, Keith. 1983. *Man and the Natural World: Changing Attitudes in England 1500–1800*. London: Allen Lane.

Thompson, Wright. n.d. "A History of Mistrust." *ESPN.com*. http://sports.espn.go.com/espn/eticket/story?page=vicksatlanta.

Tong, Rosemarie. 1993. *Feminine and Feminist Ethics*. Belmont, CA: Wadsworth.

Trans and Womyn's Action Camp. 2012. http://twac.wordpress.com/ (accessed December 31).

Tronto, Joan C. 1993. *Moral Boundaries: A Political Argument for an Ethic of Care*. New York: Routledge.

Turner, James. 1980. *Reckoning with the Beast: Animals, Pain and Humanity in the Victorian Mind*. Baltimore and London: Johns Hopkins University Press.

Twine, R. 2001. "Ma(r)king Essence—Ecofeminism and Embodiment." *Ethics & the Environment* 6 (2): 31–58.

—2010a. "Broadening the Feminism in Feminist Bioethics." In *Feminist Bioethics: At the Center, on the Margins*, J. L. Scully, L. Baldwin-Ragaven and P. Fitzpatrick (eds). Baltimore: Johns Hopkins University Press. 45–59.

—2010b. "Intersectional Disgust? Animals and (eco) feminism." *Feminism & Psychology* 20 (3): 397–406.

—2013. "Is Biotechnology Deconstructing Animal Domestication? Movements toward 'Liberation'." *Configurations* 21 (2): 135–58.

Unferth, Deb Olin. 2011. "Interview with Gary Francione." *Believer Magazine* February.

Varner, Gary. 2012. *Personhood, Ethics, and Animal Cognition: Situating Animals in Hare's Two-level Utilitarianism*. Oxford: Oxford University Press.

Villagra, Analia. 2011. "Cannibalism, Consumption, and Kinship in Animal Studies." In *Making Animal Meaning*, L. Kalof et al. (eds). East Lansing: Michigan State University Press.

Wadiwel, D. 2012. "Thinking through Race and its Connection to Critical Animal Studies." Presented at the ICAS roundtable at Minding Animals 2 Conference, July 4, Utrecht, The Netherlands.

Walker, Margaret Urban. 1989. "What Does the Different Voice Say?: Gilligan's Women and Moral Philosophy." *The Journal of Value Inquiry* 23: 123–34.

—1995. "Moral Understandings: Alternative 'Epistemology' for a Feminist Ethics." In *Justice and Care: Essential Readings in Feminist Ethics*, Virginia Held (ed.). Boulder, CO: Westview Press. First published in *Hypatia* 4, no. 2 (1989): 15–28.

—2006. *Moral Repair: Reconstructing Moral Relations after Wrongdoing.* Cambridge: Cambridge University Press.

Ward, Edith. 1892. "Review of Henry Salt, *Animal Rights*." *Shafts* 1, no. 3 (November 19).

Warren, Karen J. 1987. "Feminism and Ecology: Making Connections." *Environmental Ethics* 9 (1): 3–20.

—1990. "The Power and the Promise of Ecological Feminism." *Environmental Ethics* 12 (2): 125–46.

—(ed.). 1994. *Ecological Feminism.* New York: Routledge.

—(ed.). 1997. *Ecofeminism: Women, Culture, Nature.* Bloomington, IN: Indiana University Press.

—2000. *Ecofeminist Philosophy: A Western Perspective on What It Is and Why It Matters.* Lanham, MD: Rowman & Littlefield Publishers, Inc.

Warren, Mary Anne. 1997. *Moral Status: Obligations to Persons and Other Living Things.* Oxford: Oxford University Press.

Weil, Kari. 2010. "Shameless Freedom." *jac* 30 (3–4): 713–26.

Weil, Simone. 1968a (1941). "Classical Science and After." In *On Science, Necessity, and the Love of God*, ed. and trans. Richard Rees. London: Oxford University Press, 3–43.

—1968b. "Fragment: Foundation of a New Science." In *On Science, Necessity, and the Love of God*, ed. and trans. Richard Rees. London: Oxford University Press, 79–84.

Weiner, Russell. 2012. "Flesh." Unpublished manuscript.

Weisberg, Zipporah. 2009. "The Broken Promises of Monsters: Haraway, Animals and the Humanist Legacy." *Journal for Critical Animal Studies* 7 (2): 22–62.

White, Lynn, Jr. 1967. "The Historical Roots of Our Ecologic Crisis." *Science* 155: 3767, 1203–7.

White, R. (ed.). 2004. *Controversies in Environmental Sociology.* Cambridge: Cambridge University Press.

Wieringa, Saskia E., Evelyn Blackwood, and Abha Bhaiya (eds). 2007. *Women's Sexualities and Masculinities in a Globalizing Asia.* New York: Palgrave Macmillan.

Williams, Bernard. 1985. *Ethics and the Limits of Philosophy.* Cambridge, MA: Harvard University Press.

Williams, Terry Tempest. 1995. *Desert Quartet: An Erotic Landscape.* New York: Pantheon Books.

Winterson, Jeanette. 1992. *Written on the Body.* New York: Vintage Books.

"Women Working for Animals." *Feminists for Animal Rights Newsletter* 1:4 (Spring 1985): 2.

Woolf, Virginia. 1928 (1956). *Orlando.* New York: Harcourt Brace Jovanovich.

Wordsworth, William. 1967. "Lines Written in Early Spring" (1798). In *English Romantic Writers*, David Perkins (ed.). New York: Harcourt, Brace & World.

Wu, Peichen. 2007. "Performing Gender Along the Lesbian Continuum: The Politics of Sexual Identity in the Seitô Society." In *Women's Sexualities and Masculinities in a Globalizing Asia*, Saskia E. Wieringa, Evelyn Blackwood, and Abha Bhaiya (eds). New York: Palgrave Macmillan, 77–100.

Index

Page references in *italic* denote figures/illustrations